No. 1645
$21.95

USING
INTEGRATED
CIRCUIT
LOGIC
DEVICES

BY DELTON T. HORN

A Gift from the Estate of
Michael Hart
www.gutenberg.org

TAB TAB BOOKS Inc.

BLUE RIDGE SUMMIT, PA. 17214

FIRST EDITION

FIRST PRINTING

Copyright © 1984 by TAB BOOKS Inc.

Printed in the United States of America

Reproduction or publication of the content in any manner, without express permission of the publisher, is prohibited. No liability is assumed with respect to the use of the information herein.

Library of Congress Cataloging in Publication Data

Horn, Delton T.
    Using integrated circuit logic devices.

    Includes index.
    1. Digital integrated circuits. 2. Logic circuits.
3. Digital electronics. I. Title.
TK7874.H68 1984   621.381′73   83-24158
ISBN 0-8306-0645-9
ISBN 0-8306-1645-4 (pbk.)

Contents

Introduction

Digital electronics devices are becoming more and more important in today's technological world. They are even turning up in products that were previously the sole domain of linear/analog components (such as television and radio receivers). A modern technician or hobbyist needs a firm working knowledge of digital electronics to keep up with the rapid technological advancements throughout the electronics industry.

Unfortunately, many people working with electronics don't understand or feel uncomfortable with digital techniques. Many fear that the complexities of digital electronics may be beyond them. Actually, digital electronics is a field that is not difficult to understand if you first comprehend a few basic principles.

The basic principles of digital electronics are presented in this book through discussions of theory, and with practical circuits and projects. When you finish this book, you should be ready to enter the exciting world of modern digital technology with confidence. At long last you'll know just what is happening inside those mysterious little components known as digital integrated circuits. You will be able to analyze, troubleshoot, and even design circuits using up-to-date digital electronics techniques. Best of all, you'll learn just how easy it really is!

Chapter 1

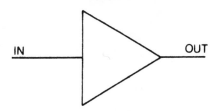

Introduction to Digital Logic

Virtually all electronic circuits fit into one of two broad classes—analog or digital. A few circuits are a hybrid of the two, but the majority are either one or the other. Until relatively recently most electronics technicians were more familiar with analog type circuits. The analog category includes devices like amplifiers, power supplies, oscillators, radio transmitters and receivers, and so forth. However, thanks to improved IC technology, and the rapid development of computers and related devices, digital circuits are growing in importance. They are now so commonly used in so many applications that anybody who even dabbles in electronics should have a firm foundation in digital circuitry principles.

DIGITAL VERSUS ANALOG

The first thing to consider is how digital and analog circuits differ. In an analog circuit, the output is a direct analog of the input, although it may be changed in some way. For instance, in an amplifier, the output is essentially the same as the input, except it is at a higher level.

Analog circuits are sometimes called linear circuits, because the output can be graphed linearly (see Fig. 1-1). The output voltage (or current) can change over a continuous range. There are a theoretically infinite number of possible inputs. Of course, in practice, the differences may be too small to measure.

A digital circuit, on the other hand, is far more limited in the

1

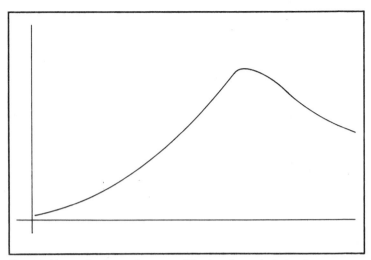

Fig. 1-1. Linear circuits produce smoothly changing output signals.

signals it can handle. Each digital input and output can deal with only two possible states. Either the voltage is low, or the voltage is high. There is no possible middle ground. In essence, a digital circuit deals only with "yes/no" conditions, while an analog circuit can handle a wide range of "maybe's."

At first glance, it might seem that this would make digital circuits far too limited to be of any practical value. In truth, however, nothing could be further from the reality. As the following chapters will demonstrate, digital circuits are extremely useful and highly versatile. In fact, many applications can be accomplished digitally that would be impractically difficult, or even impossible, to achieve by analog means.

Digital circuits are ideal for electronic switching, computers and other programmable devices, long distance signal transmission, information storage, and many other important applications. There is some overlap between applications for digital and analog circuits, but in many cases, the digital approach offers a number of significant advantages. Let's look at signal transmission for an example. A basic block diagram is shown in Fig. 1-2. The transmitter originates a signal which is sent along a transmission line to a receiver. Ideally, the received signal should be identical to the transmitted signal.

Unfortunately, in the real world, a number of things can happen to the signal as it passes through the transmission line. For one thing, some of the original signal will be lost along the length of the line. If the transmission line is very long, preamplifiers may have to be used or the received signal may be too weak to be useful.

Another problem is that any length of wire can act as an antenna, picking up unwanted signals from radio transmitters and/or ac power lines. This noise is added to the original signal and sent right along to the receiver. Shielded cables or twisted pair wiring can help reduce (but not always completely eliminate) noise pick-up, but they can add to the cost and bulk of the transmission line. Coaxial cable must be of the correct impedance or distortion and excessive signal loss will result. The shielding can also set up stray capacitances along the transmission line that could be a problem in some instances. Moreover, certain frequencies may be attenuated more than others, distorting the received waveshape even more.

These combined effects can often render a received analog signal completely useless. Figure 1-3 illustrates a particularly bad case of a transmitted and received signal. There is no way for the receiver to distinguish the noise from the desired signal, so it is not likely to be able to recover the original signal very well, if at all. (Of course, most analog transmissions aren't this bad.)

If digital signals are transmitted, however, we only have to be able to recognize whether the signal is at a low or high level. Noise in between the two levels, or beyond the extremes will not matter

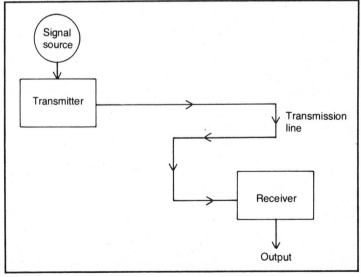

Fig. 1-2. Typical signal transmission/reception system.

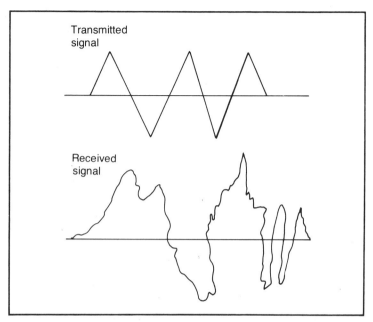

Fig. 1-3. When analog signals are transmitted, they may be badly distorted by the time they are received.

unless it is severe enough to confuse the two relevant levels. An example of a transmitted and received digital signal is shown in Fig. 1-4. Notice that while the distortion is as bad as in the analog example (Fig. 1-3), it is much easier to recognize the original signal and ignore the noise.

Some digital transmission schemes transmit two distinct frequencies to represent the logical high and low levels. The receiver can ignore all other frequencies. A significant amount of noise is not likely to occur at the precise frequencies used for encoding.

All this boils down to the fact that digital transmission is often more reliable than analog transmission, especially in very noisy environments. This is not to imply, however, that digital signal transmission is perfect, or in any way foolproof. Errors can occur. Fortunately, many safeguards can be set up, and these would not be possible in an analog system. Digital signal transmission will be covered in greater depth in a later chapter.

Digital circuitry is not always the best approach to every electronic circuit. In many cases, an analog system is required. But digital circuits are extremely versatile, and the electronics technician or hobbyist must understand how they work, or he will be completely in the dark when faced with much of today's (and tomorrow's!) electronic equipment.

THE IC

It is certainly possible to build digital circuits with discrete components such as transistors, or even tubes. But if the circuitry is at all complex, the project can become hopelessly unwieldy and expensive. Digital circuitry didn't really begin to come into its own until the development of the *integrated circuit,* or IC.

In the early days of electronics, all circuits were constructed around vacuum tubes. While tubes were a technological miracle in their time, they suffered from a number of disadvantages. For one thing, they were relatively expensive, and tended to have a fairly short lifetime. Tubes burn out eventually, like light bulbs. Moreover, their operating characteristics tend to change as they age.

Probably the biggest limiting factor to tube technology was bulk. Tubes are fairly large devices. In addition, they give off a great deal of heat, which means components had to be spaced widely apart to avoid thermal damage. Tubes also require a secondary power supply to heat their filaments. All of these problems tended to limit the practical limits of tube based circuitry. Many complex circuits would simply be too large, expensive and generally unwieldy to be practical.

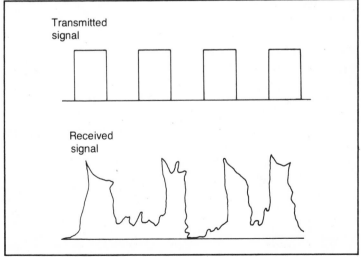

Fig. 1-4. Transmission distortion is of less significance in a digital system.

5

For example, modern computers would be unthinkable if constructed from vacuum tubes. They would be theoretically possible, but completely infeasible in actual practice. True, the earliest computers were tube based, but they were fabulously expensive and gigantic monsters, consuming enormous amounts of power. They were also extremely limited in their capabilities as compared to today's computers. An early office-filling tube computer probably wasn't as powerful a computing device as many modern hand-held programmable calculators.

Then, in the fifties, the transistor was developed. This device could accomplish just about anything a vacuum tube could do, but in a fraction of the space, at a fraction of the power, and dissipating a fraction of the waste heat of a comparable tube circuit. Transistors take advantage of the unique properties of certain crystaline substances such as silicon and germanium. Most substances are either conductors (will allow electric current to pass through them fairly easily, like copper or silver) or insulators (will tend to block the passage of electric current through them, like glass or rubber). A few elements, however, don't quite fit into either category. These materials are called semiconductors, and they are the secret behind both transistors and integrated circuits.

In electronic components, such as transistors, the semiconductor crystal is doped with a small amount of another substance, called an impurity. Some impurities will result in a crystal with an excess of electrons, producing a slab of N-type material. Other materials result in a semiconductor with a shortage of electrons, or a P-type material.

N-type and P-type slabs are sandwiched together in various combinations to create electronic components. For instance, a bipolar transistor is made up of three slabs. The two outer slabs are of one type substance, while the middle slab (which is much thinner than the other two) is of the other type. An NPN transistor is illustrated in Fig. 1-5, while Fig. 1-6 shows a PNP transistor.

It is not the purpose of this book to explain semiconductor action and transistor theory in detail. Fortunately, there is no vital reason for the experimenter to know the details of how transistors, diodes, etc., are manufactured. Many fine books on the principles of transistors are available for interested readers.

Transistors offered a number of significant improvements over tubes. They tended to be less expensive to manufacture, they ran cooler, they were far more reliable, and they allowed circuits to be constructed in far smaller spaces. An electronic device that might

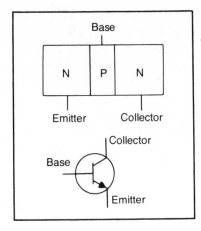

Fig. 1-5. An NPN transistor is constructed of two slabs of N-type semiconductor sandwiching a thinner slab of P-type material.

fill a good sized room if constructed of vacuum tubes, might be compressed into the size of a desk in the transistor version.

Once the idea of miniaturization turned up, circuit designers began looking for ways to miniaturize circuits still further. Printed circuit boards came into wide-spread use, and many manufacturers started working with modules.

A circuit module is simply several components wired together in a standard configuration (say, an amplifier, or an oscillator circuit), compressed together and encapsulated. Appropriate leads are brought out from the module to connect with external components. The use of modules cut down circuit assembly time, and could often reduce space requirements by as much as 25%.

Fig. 1-6. A PNP transistor is constructed of two slabs of P-type semiconductor sandwiching a thinner slab of N-type material.

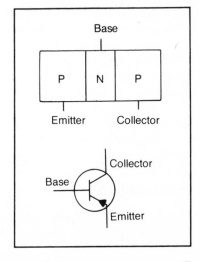

Another advantage of using these circuit modules was that they simplified the task of circuit design. Each module could be used as a "black box"—if you put in signal "A," the module will put out signal "B." The system designer doesn't have to worry about designing each individual stage—he only has to put the stages together in the appropriate way.

While circuit modules were something of an improvement, they still left a lot to be desired. They were still relatively bulky, and module assembly costs had to be incorporated into the price. While convenient under some circumstances, the idea simply wasn't too practical on a large scale.

Then along came the integrated circuit, or IC. This was actually a further development in transistor technology. Multiple transistors, diodes, resistors, and capacitors can be etched onto a single slab of semiconductor material. The earliest ICs were not too impressive by today's standards, but they were a miracle of their time. Early devices could replace a handful of discrete transistors and/or resistors in a circuit. Replacing, say, three transistors and half a dozen resistors with a single component that measured less than 1-inch by ¼-inch allowed significant reduction in circuit size.

At first, integrated circuits were quite expensive. Early prototypes often cost more than $100 apiece. The costs incurred in research and development had to be covered. The initial high cost limited the use of integrated circuits to very special projects by the government or large industrial interests. It only made sense to pay the added expense in cases where small size was a critical factor, such as in missiles and satellites.

As more and more integrated circuits were sold, the development costs ceased to be an important factor in pricing. Prices dropped steeply, and today many integrated circuit devices are readily available for well under $1 apiece. The lower the price, the more units were sold, which allowed the price to drop further. Today ICs are used in the majority of electronic circuits. Quite often using an integrated circuit actually works out cheaper than using discrete components. This is especially true for complex devices.

The earliest ICs were very simple devices, as mentioned earlier. As the technology improved, more and more components could be etched onto a single silicon chip, allowing for more and more complex devices. Integrated circuits are often identified by their level of complexity. There is a great deal of overlap between the categories—the borders are not firmly defined. Still, the terms are useful in distinguishing high level from low level ICs.

Modern integrated circuits are usually divided into the four broad classes listed below:

- ☐ SSI   Small Scale Integration (simple devices)
- ☐ MSI   Medium Scale Integration (moderately complex devices)
- ☐ LSI   Large Scale Integration (complex devices)
- ☐ VLSI   Very Large Scale Integration (extremely complex devices)

Some modern integrated circuits may include up to 5000 or 6000 etched components. Earlier I mentioned a vacuum-tube circuit that filled a room and its transistorized equivalent that was the size of a desk. An integrated circuit version of the same device might well be able to sit on your fingertip.

In fact, the practical limit to integration seems to be not so much the number of components to be etched, but the number of external input and output pins required. Forty or fifty pins seems to be about the maximum for practical devices. More than that would result in severe handling problems.

The actual integrated circuit is not much bigger than the head of a pin. This is too small for human beings to work with, so it is usually enclosed in a somewhat larger package, with pins extended out for external connections. Three package types are the most common.

Some ICs are contained in round cans, similar to some transistors. These round packages have from three to about 12 leads in a circular pattern. The pins are numbered in a clockwise fashion for convenient identification. A small tab on the case is used to distinguish pin number 1. See Fig. 1-7.

Another common type of IC packaging is the flat pack, as shown in Fig. 1-8. The pins extend outward from the sides and are soldered directly to the traces of a printed circuit board. Flat pack ICs are ideal for automated circuit assembly, but they are not too practical for manual assembly.

The most popular IC packaging scheme is the DIP, or *Dual In-line Package*. This type of packaging is illustrated in Fig. 1-9. A notch is used to identify the front of the device. Be careful when working with DIP ICs—some sources count the pins from the top, while a few count from the bottom. In this book we will adhere to the convention of counting the pins while looking down on the device from above.

DIP ICs are easy to handle, and convenient sockets are available to reduce the risk of damage while soldering. Of course, the

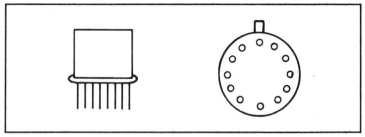

Fig. 1-7. Some integrated circuits are enclosed in round cans.

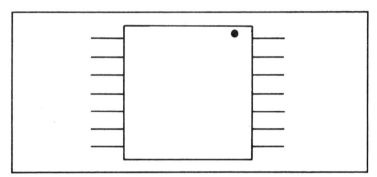

Fig. 1-8. Some integrated circuits are enclosed in flat packs.

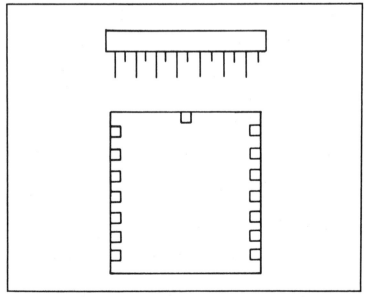

Fig. 1-9. Most experimenters use integrated circuits in the DIP (dual inline pin) package.

use of sockets also makes replacement a simple matter if it should ever prove necessary. DIP ICs come in a wide variety of sizes. Eight pin, fourteen pin, and sixteen pin devices are the most plentiful, but other sizes (24 pin, 28 pin, and even 40 pin) are not uncommon. The number of pins is always an even number.

Some devices require an odd number of pins, or don't work out to one of the standard DIP sizes. Rather than manufacture a number of odd ball packages, one of the standard sizes is usually employed, leaving some of the pins unconnected internally to the IC chip. This is generally represented on pinout diagrams as "N.C."

BINARY ARITHMETIC

This book is on digital integrated circuits. The name "digital" implies the use of numbers or "digits." How does an electronic circuit count and perform mathematical operations? Before answering that, let's first review how we handle these tasks.

The Ordinary Decimal System

We are used to numbers in the decimal system. The term simply refers to counting by tens. In other words, each digit of a number may be one of ten possible values (0, 1, 2, 3, 4, 5, 6, 7, 8, or 9). If we need to express a number larger than nine, we add another digit to the left of the first. For example, 37 means 3 tens and 7 ones. For still larger numbers, a third column is added. The base value of this column is ten squared (10^2) or hundreds. The next column is 10^3 or thousands, then 10^4 or ten-thousands, and so forth.

In decimal addition it is often necessary to carry part of a column's result over into the next column. For example, if we are adding 58 and 23 we first add the ones columns to get $8 + 3 = 11$. We put 1 into the ones column and carry 1 over into the tens column. Adding the tens we now have $5 + 2 + 1$ (carried over from the ones column) $= 8$. Putting the columns back together, we have the complete solution—$58 + 23 = 81$.

Similarly, in subtraction, we sometimes have to borrow from the next highest column. For instance, let's try $64 - 35$. First we work with the ones column, or $4 - 5$. Obviously 5 is too big to be subtracted from 4, so we borrow 1 from the tens column, giving us $14 - 5 = 9$. Next we subtract the tens. Remember that we borrowed one ten, so we now have $(6 - 1 \text{ (borrowed)}) - 3 = 5 - 3 = 2$. Putting the columns back together for the full solution we find that $64 - 35 = 29$.

This is all quite elementary, of course. Most of us can breeze through these steps without even consciously thinking about them. An understanding of how the decimal system works is necessary for understanding other number systems.

There is nothing unique about the decimal number system. We use the base 10 because we happen to have ten figures. But numbering systems can be based on other numbers. The ones we are concerned with are the binary system (base 2), the octal system (base 8) and the hexadecimal system (base 16). These alternative number systems will be discussed in the next few pages.

Counting by Twos

The binary system is an extremely important concept in digital electronics. This numbering system has a base of two. This means there are only two possible digits—0 and 1. Any number greater than 1 must be carried over into the next highest column.

The first column, of course is the ones column. Immediately to the left of the ones column is the twos column. Next we have 2^2 (fours), 2^3 (eights), 2^4 (sixteens), 2^5 (thirty-twos), and so forth.

The binary system may sound confusing, but it really isn't all that much different from the more familiar decimal system. For instance, the decimal number 3728 breaks down into 3 thousands, 7 hundreds, 2 tens, and 8 ones. Similarly, the binary number 110110 breaks down to 1 thirty-two, 1 sixteen, 0 eights, 1 four, 1 two, and 0 ones. In other words, binary 110110 is equal to decimal 32 + 16 + 4 + 2, or 54. Binary and decimal numbers are compared in Table 1-1.

Sometimes, a subscript is added to numbers to identify the base. That is, decimal 3728 might be written as 3728_{10} and binary 110110 could be written as 110110_2. This helps prevent confusion. After all, 110110 could be a decimal number (one-hundred and ten thousand, one-hundred and ten).

Binary addition works the same way as decimal addition. Let's add binary 1001 and 1100. First, we add the ones column—1 + 0 = 1—simple enough. In the twos column we have 0 + 0 = 0. Moving on to the fours column we have 0 + 1 = 1. Finally, in the eights column we have 1 + 1. Of course this works out to a value of two, but the highest possible digit in the binary system is 1. We have to put a 0 into the eights column and carry over a 1 into the next highest column—the sixteens column. In this case, there is nothing else in the sixteens column, so that column simply has a value of 1. Putting all of the columns back together we find that 1001 + 1100 = 10101. That looks okay but is it the right answer?

We can check our results by converting the binary numbers

Table 1-1. Binary and Decimal Numbers.

binary	decimal	binary	decimal
000000	0	100100	36
000001	1	100101	37
000010	2	100110	38
000011	3	100111	39
000100	4	101000	40
000101	5	101001	41
000110	6	101010	42
000111	7	101011	43
001000	8	101100	44
001001	9	101101	45
001010	10	101110	46
001011	11	101111	47
001100	12	110000	48
001101	13	110001	49
001110	14	110010	50
001111	15	110011	51
010000	16	110100	52
010001	17	110101	53
010010	18	110110	54
010011	19	110111	55
010100	20	111000	56
010101	21	111001	57
010110	22	111010	58
010111	23	111011	59
011000	24	111100	60
011001	25	111101	61
011010	26	111110	62
011011	27	111111	63
011100	28	1000000	64
011101	29	1000001	65
011110	30	1000010	66
011111	31	1000011	67
100000	32	1000100	68
100001	33	1000101	69
100010	34	1000110	70
100011	35		And so forth . . .

into their decimal equivalents. Binary 1001 is equal to $(1 \times 8) + (0 \times 4) + (0 \times 2) + (1 \times 1)$, or decimal 9. Binary 1100 has a decimal value of $(1 \times 8) + (1 \times 4) + (0 \times 2) + (0 \times 1)$, or 12. Finally, binary 10101 equals decimal $(1 \times 16) + (0 \times 8) + (1 \times 4) + (0 \times 2) + (1 \times 1)$, or 21. Adding decimal 9 to decimal 12, we also get decimal 21, so our solution was correct.

While you should understand the principles of binary numbers, you will rarely have to work directly with them. That is the job of digital circuits.

Boolean Algebra

Addition, subtraction, multiplication, and division are the most

13

A	$\overline{A}$
0	1
1	0

Table 1-2. The Boolean Algebra NOT Function Can Be Demonstrated in a Truth Table.

familiar mathematical operations. In the binary system, and digital electronics, there is a more important set of mathematical operations known as Boolean algebra. This may sound complicated at first, but once you are more familiar with it, you should find it fairly easy. A knowledge of Boolean algebra is virtually essential for working with digital logic ICs.

There are three basic functions in Boolean algebra. They are NOT, AND, and OR. NOT is simply a matter of reversing the value of a digit. Since there are only two possible digit values (0 and 1), if a digit is NOT 0, it must be 1. Conversely, if it is NOT 1, it must be 0. Certainly there is nothing complicated there.

Table 1-3. The Boolean Algebra AND Function Can Be Demonstrated in a Truth Table.

A	B	A•B
0	0	0
0	1	0
1	0	0
1	1	1

The NOT operation is usually indicated by a bar over the number to be reversed. NOT A is the same as $\overline{A}$. The NOT operation is summarized in Table 1-2. This is called a truth table. It is a handy way to summarize all possible input and output combinations for a Boolean algebra operation.

Only a single number is involved in the NOT operation. AND and OR require at least two numbers to be combined. The AND operation is usually written as A•B for A AND B. The truth table for the AND operation is given in Table 1-3.

The OR operation is generally notated as A + B for A OR B. The OR truth table is shown in Table 1-4. You will learn about the significance of these basic Boolean algebra operations as we study the various logic gates in the following chapters.

A	B	A+B
0	0	0
0	1	1
1	0	1
1	1	1

Table 1-4. The Boolean Algebra OR Function Can Be Demonstrated in a Truth Table.

The Octal System

While the binary system is convenient for digital electronic circuits, it is awkward, at best, for humans. It is difficult to know the value of a binary number like 110100101 at a glance.

The octal system is a popular compromise for people who have to work with large binary numbers. This is a number system with a base of eight. A binary number can easily be converted into an octal number simply by breaking it up into groups of three digits. A three digit binary number may take any of eight different values:

$$000 = 0$$
$$001 = 1$$
$$010 = 2$$
$$011 = 3$$
$$100 = 4$$
$$101 = 5$$
$$110 = 6$$
$$111 = 7$$

(There is no digit 8 or 9 in the octal system.)

Let's break down binary 110100101 into the octal equivalent. First, we start at the right-most digit (ones column) and group the digits into sets of three:

$$110 \quad 100 \quad 101$$

Binary 110 is equal to octal 6, binary 100 is octal 4, and binary 101 equals octal 5. Putting the number back together we find binary 110 100 101 equals octal 645, which is far easier to remember and copy without making a mistake.

Octal numbers can be converted to the decimal system, just as binary numbers can. Octal 645 equals decimal $(6 \times 8^2) + (4 \times 8) + (5 \times 1) = (6 \times 64) + (4 \times 8) + (5 \times 1) = 384 + 32 + 5 = 421$.

The Hexadecimal System

Another popular number system is the hexadecimal (base 16 system). It can be derived from the binary system in a similar way to the octal system. For hexadecimal numbers, however, the binary digits are grouped into sets of four, giving sixteen possible values:

$$0000 = 0$$
$$0001 = 1$$

$$0010 = 2$$
$$0011 = 3$$
$$0100 = 4$$
$$0101 = 5$$
$$0110 = 6$$
$$0111 = 7$$
$$1000 = 8$$
$$1001 = 9$$
$$1010 = 10$$
$$1011 = 11$$
$$1100 = 12$$
$$1101 = 13$$
$$1110 = 14$$
$$1111 = 15$$

Each group of four binary digits is replaced by a single hexadecimal digit. Unfortunately, we have no single digits to represent values of 10 through 15. The letters A through F are used to notate these values in the hexadecimal system. Hexadecimal B, for example, is equal to decimal 11.

Returning to the earlier example we used to convert from binary to octal, let's now convert binary 110100101 into its hexadecimal equivalent. Breaking the number up into groups of four digits, we get:

$$1 \quad 1010 \quad 0101$$

Leading zeroes may be added to the highest value group to complete the set of four. This does not affect the value of the number in any way:

$$0001 \quad 1010 \quad 0101$$

The hexadecimal equivalent for binary 0001 is just 1, of course. Binary 1010 is equal to hexadecimal A (decimal 10), and binary 0101 is the same as hexadecimal 5. Putting the digits back together, we have:

$$0001 \quad 1010 \quad 0101_2 = 1A5_{16} = 645_8 = 421_{10}.$$

The four basic number systems are compared in Table 1-5.

Table 1-5. Here Is a Comparison of Binary, Octal, Hexadecimal and Decimal Numbers.

Decimal	Binary	Octal	Hexadecimal
0	0000 0000	0	0
1	0000 0001	1	1
2	0000 0010	2	2
3	0000 0011	3	3
4	0000 0100	4	4
5	0000 0101	5	5
6	0000 0110	6	6
7	0000 0111	7	7
8	0000 1000	10	8
9	0000 1001	11	9
10	0000 1010	12	A
11	0000 1011	13	B
12	0000 1100	14	C
13	0000 1101	15	D
14	0000 1110	16	E
15	0000 1111	17	F
16	0001 0000	20	10
17	0001 0001	21	11
18	0001 0010	22	12
19	0001 0011	23	13
20	0001 0100	24	14
21	0001 0101	25	15
22	0001 0110	26	16
23	0001 0111	27	17
24	0001 1000	30	18
25	0001 1001	31	19
26	0001 1010	32	1A
27	0001 1011	33	1B
28	0001 1100	34	1C
29	0001 1101	35	1D
30	0001 1110	36	1E
31	0001 1111	37	1F
32	0010 0000	40	20
33	0010 0001	41	21
34	0010 0010	42	22
35	0010 0011	43	23
36	0010 0100	44	24
37	0010 0101	45	25
38	0010 0110	46	26
39	0010 0111	47	27
40	0010 1000	50	28
41	0010 1001	51	29
42	0010 1010	52	2A
43	0010 1011	53	2B
44	0010 1100	54	2C
45	0010 1101	55	2D
46	0010 1110	56	2E
47	0010 1111	57	2F
48	0011 0000	60	30

Why Binary?

You may be wondering why we bother with all these different number systems. Why not just stick with the good old decimal system? To electronically simulate the decimal systems, we would need some way to generate and recognize ten different, distinct, and unambiguous values—one for each possible digit value. This would result in some very complex circuitry. However, the binary system requires only two possible values. This can be electronically encoded with a simple switch—off represents 0, and on stands for 1. For the octal and hexadecimal systems we can combine binary switches into groups of three or four.

Most digital electronics circuits use electronic switches (such as transistors, or special circuits known as "gates") rather than mechanical switches, but the principle is the same. Binary 0 is represented by a very low voltage (ground, or close to it), while binary 1 is indicated by a higher voltage.

Hundreds, or even thousands of electronic switches can be combined to perform almost any mathematical operation in the binary system. The binary system is awkward for us to work with because even moderate values result in large numbers that can easily be confused. But in electronics, the binary system is far, far easier to implement than the decimal system.

Chapter 2

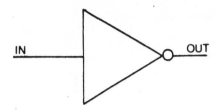

One-Input Digital Gates

Digital electronic circuits are generally made of "building block" subcircuits known as gates. One or more digital gates of a specific type are contained in SSI (Small-Scale Integration) integrated circuits. LSI (Large-Scale Integration) ICs may contain hundreds, or even thousands, of digital gates.

A digital gate accepts one or more inputs (each input is fed a binary digit—0 or 1) and produces one or more binary digits as output. The value of the output is dependent upon the values presented to the inputs. If the input conditions are known, the output will be predictable. The input and output conditions may be summarized in a truth table like those used to demonstrate Boolean algebra in Chapter 1. In fact, some digital gates perform Boolean algebra operations on the inputted numbers.

FAN-IN AND FAN-OUT

The output of one digital gate may be used as the input of another gate. How many additional gates may be driven from a single source is dependent upon the fan-in and fan-out characteristics of the devices in question. The fan-in of a gate is simply its input voltage and current requirements. For convenience the fan-in is usually expressed in terms of a standard unit. The standard unit will vary for different logic families (see Chapter 12).

We will use the logic family known as TTL for an example. Most simple TTL gates have a fan-in of one standard unit. Some

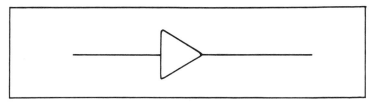

Fig. 2-1. A digital buffer is represented by this symbol.

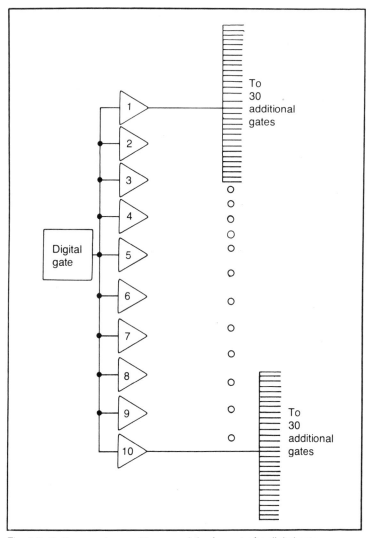

Fig. 2-2. Buffers can be used to expand the fan-out of a digital gate.

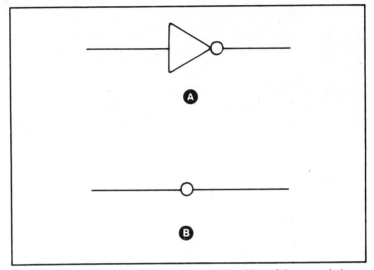

Fig. 2-3. A digital inverter may be represented by either of these symbols.

have a higher fan-in, and a few might have slightly lower requirements.

The fan-out of a digital gate is just the opposite of the fan-in. Fan-out is how many additional gates can be simultaneously driven by a single output. Fan-out is expressed in the same standard units as the fan-in. If you exceed the fan-out capabilities, the circuit will not function correctly or reliably. The fan-out for most TTL gates is about ten standard units. This means, each gate may drive about ten additional gates, assuming the additional gates have a fan-in of one.

In simple digital circuits, fan-in and fan-out often aren't of much concern, since only a few gates are used so the limits are not exceeded. However, more complex digital circuits frequently run into fan-in/fan-out problems that the designer must account for. When large numbers of gates are being used, it is essential to check

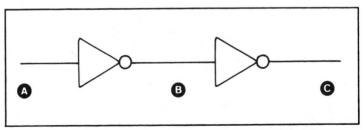

Fig. 2-4. Two inverters can be connected in series to simulate a noninverting buffer.

21

Input	Output
0	0
1	1

Table 2-1. The Truth Table for a Digital Buffer is Very Simple.

the fan-in and fan-out requirements to avoid trouble. Fan-in and fan-out specifications are also of importance when two or more logic families are used together. This topic will be discussed in Chapter 12.

THE BUFFER

Undoubtably, the simplest digital gate is the buffer. This device has a single input and a single output. Its schematic symbol is shown in Fig. 2-1. The truth table for a buffer is shown in Table 2-1.

If you have looked at the truth table, you may be wondering what a buffer is good for. The output is always the same as the input, which may seem rather pointless. The purpose of the buffer was suggested in the previous section of this chapter. A buffer is essentially a fan-out amplifier. That is, it allows a single output to drive more inputs.

Most buffers have a fan-in of one. A typical TTL buffer has a fan-out of thirty. A single digital gate with a fan-out of ten could drive ten buffers each driving thirty additional gates. This would allow a single output to simultaneously drive up to 300 additional digital gates. This idea is illustrated in Fig. 2-2.

The buffer is handy to have when working with complex digital circuits. It increases the fan-out without interfering with the logic in anyway. The most common form of buffer is an IC called a hex buffer. As the name states, this is an IC containing six separate buffer gates. Some other types of digital gates include buffers within the same integrated circuit.

THE INVERTER

Another single input/single output digital gate is the inverter. The output is inverted, or made the opposite of the input. The truth

Table 2-2. The Truth Table for an Inverter is the Same as the One for the Boolean Algebra NOT Function.

Input	Output
0	1
1	0

**Table 2-3. This Truth Table Demonstrates How
Two Inverters in Series Can Simulate a Noninverting Buffer (See Fig. 2-4).**

A (input)	B	C (output)
0	1	0
1	0	1

table for an inverter is shown in Table 2-2. The schematic symbol
for an inverter is shown in Fig. 2-3A. Sometimes the triangle is
eliminated, and just a circle is used to represent an inverter, as in
Fig. 2-3B.

Inverters often have relatively large fan-outs, so they can be
used as buffers. Two can be connected in series, so the final output
is the same as the input. This is demonstrated in Fig. 2-4 and Table
2-3.

Six inverters are generally available on a single IC called a hex
inverter. Inverters are most useful in combinations with other
gates.

Chapter 3

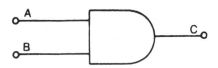

The AND Gate

Now we will turn to a more complex gate, called the AND gate. The output of an AND gate is a 1 if, and only if, all of the inputs are 1's. If one or more of the inputs is a 0, then the output will be a 0. Obviously an AND gate requires a minimum of two inputs. The truth table for a two-input AND gate is shown in Table 3-1. Does it look familiar? It should! An AND gate performs a Boolean algebra AND operation.

To get a better idea of how an AND gate works, take a look at the simple circuit shown in Fig. 3-1. The lamp will light only if both switches are closed (logic 1). If one, or both of the switches is open (logic 0), the current path will remain broken and the lamp will not light. The same principle can be extended to include as many switches (digital inputs) as you like. See Fig. 3-2.

The mechanical switches in Figs. 3-1 and 3-2 can be replaced with semiconductors that can act as electronic switches. Figure 3-3, for example, shows a simple two-input AND gate using diodes. If either (or both) of the inputs is at a logic 0, the cathode of the appropriate diode will be effectively grounded. This forward biases the diode, allowing it to conduct. A circuit path is created from $+V_{CC}$, through R1, and one of the diodes to ground. A forward biased diode has a very low resistance, so a very low voltage will appear at the output. This voltage will be so close to zero, the difference won't much matter, so the output is logic 0.

If logic 1's are fed to both inputs (i.e., the cathodes are con-

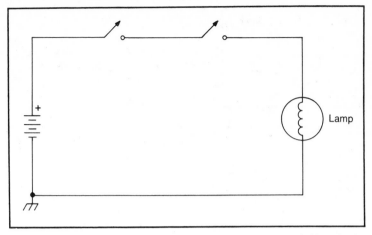

Fig. 3-1. The AND function can be simulated with this simple switching circuit.

nected to V_{CC}), the diodes will be reverse biased, and will present a very high resistance. This allows most of the V_{CC} voltage to appear at the output for a logic 1. Notice how if either diode is forward biased, the output will drop to zero.

Transistor switches can also be used to create an AND gate. A simplified typical circuit is illustrated in Fig. 3-4. This transistorized AND gate is far superior to the diode version of Fig. 3-3. There is a great deal of isolation between the inputs and the output, decreasing loading problems.

The transistors are used as electronic switches, and are either cut-off or saturated. Transistors Q1 and Q2 are the actual switches.

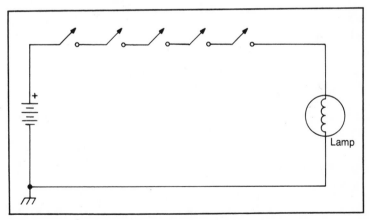

Fig. 3-2. Any number of switches may be used in an AND circuit.

Inputs		Output
A	B	C
0	0	0
0	1	0
1	0	0
1	1	1

Table 3-1. The Truth Table for an AND Gate Is the Same as One for the Boolean Algebra AND Function.

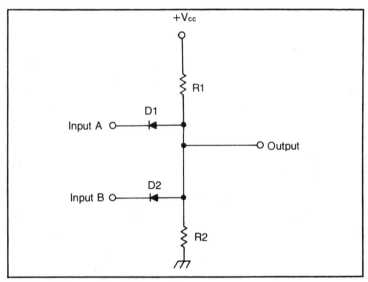

Fig. 3-3. Diodes can be used as switches to create an AND gate.

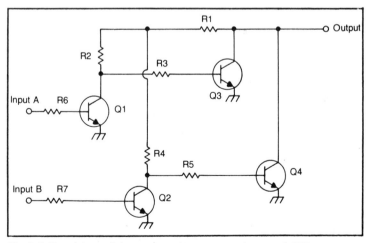

Fig. 3-4. Transistor switches are used to create an improved AND gate.

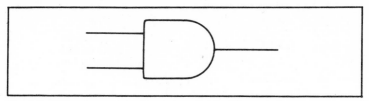

Fig. 3-5. AND gates are usually represented by this symbol.

However, their outputs are inverted 180° from their inputs so transistors Q3 and Q4 are used to re-invert the signals back to their original polarity.

Integrated circuits containing four two-input AND gates are readily available. The schematic symbol for a two-input AND gate is shown in Fig. 3-5.

MULTIPLE-INPUT AND GATES

Most standard AND gates have two inputs and a single output.

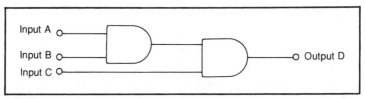

Fig. 3-6. Two AND gates can be cascaded to perform a three-input AND function.

In some circuits, however, it may be necessary to perform an AND operation on three or more simultaneous inputs. One way to accomplish this would be to cascade two or more two-input AND gates, as shown in Fig. 3-6. Multiple input AND gates are also available in IC form. The schematic symbol is the same as for the standard two-input AND gate, except for the addition of more input lines. A schematic symbol for a three-input AND gate is shown in Fig. 3-7.

Multiple input AND gates work just like the standard two-input version. The output is a logic 1 if, and only if, all the inputs are at

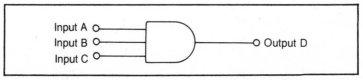

Fig. 3-7. A three-input AND gate may be represented with this symbol.

Inputs			Output
A	B	C	D
0	0	0	0
0	0	1	0
0	1	0	0
0	1	1	0
1	0	0	0
1	0	1	0
1	1	0	0
1	1	1	1

Table 3-2. The Truth Table for a Three-Input AND Gate Is Simply an Extension of the Two-Input Version.

logic 1. If any one (or more) of the inputs is at logic 0, the output of the gate will be a logic 0. The truth table for a three-input AND gate is shown in Table 3-2. Notice its similarity to the truth table for the two-input AND gate, which was given in Table 3-1. AND gates with up to eight inputs are available in integrated circuit form.

NAND GATES

Gates can be combined in various ways to create new types of gates. We have already seen this when we cascaded two two-input AND gates to create a three-input AND gate (Fig. 3-6) and two inverters to create a noninverting buffer (Fig. 2-4). Unlike gates can also be combined.

For example, look at the combination of an inverter and an AND gate shown in Fig. 3-8. Here one of the inputs to the AND gate is inverted before the AND operation. To determine how this combination of gates will work, we can draw up a truth table, as shown in Table 3-3. The output is a logic 1 if, and only if, B and $\overline{A}$ (NOT A) are both logic 1's. In other words, A must be at logic 0 and B must be at logic 1 to produce a logic 1 at the output. Any other combination will result in an output of logic 0.

In some schematic diagrams, the inverter may be represented by just a small circle, as mentioned in Chapter 2. Figure 3-9

Table 3-3. This Truth Table Shows the Results of Combining an Inverter and an AND Gate in the Manner Illustrated in Fig. 3-8.

Inputs			Output
A	B	$\overline{A}$	
0	0	1	0
0	1	1	1
1	0	0	0
1	1	0	0

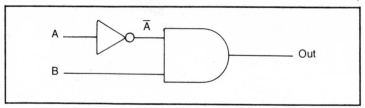

Fig. 3-8. An AND gate and an inverter can be combined to create different logic functions.

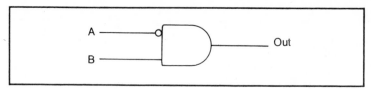

Fig. 3-9. The AND gate/inverter combination of Fig. 3-8 can also be represented with this symbol.

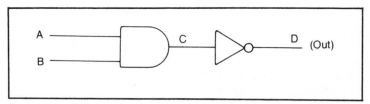

Fig. 3-10. An inverter can also be used to reverse the output of an AND gate.

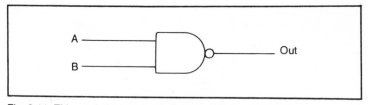

Fig. 3-11. This symbol is used to indicate a NAND gate.

illustrates this method of diagramming. This circuit is functionally identical to the one shown in Fig. 3-8.

Figure 3-10 shows another combination of a two-input AND gate and an inverter. Here the output of the AND gate is inverted before the final output. Study the truth table in Table 3-4. Notice how the output is the exact opposite of a regular AND gate. The output is a logic 1 as long as at least one of the inputs is a logic 0. The output is a logic 0 if, and only if, both inputs are logic 1's. In other words, the output is a logic 1 if A AND B are NOT 1's. This function

29

**Table 3-4. This Truth Table Demonstrates that
the NAND Function Is the Exact Opposite of the AND Function.**

Inputs			Output
A	B	C	D
0	0	0	1
0	1	0	1
1	0	0	1
1	1	1	0

is sometimes called NOT AND. This is usually shortened down to NAND.

NAND gates are extremely useful and widely available. In fact, since the input stage of a transistor AND gate automatically inverts (refer back to Fig. 3-4), a single circuit NAND gate is actually simpler and less expensive than an AND gate, which requires additional transistors to re-invert the output back to its original polarity.

Many NAND gates are available in IC form. In fact, in some circuits a NAND gate and an inverter are used to perform the AND function. The schematic symbol for a NAND gate is shown in Fig. 3-11. Notice that this symbol may represent a single NAND gate or a separate AND gate/inverter combination. They are functionally identical. NAND gates are more frequently encountered in modern circuits than AND gates because their lower parts count makes NAND ICs somewhat less expensive than comparable AND gate devices.

Chapter 4

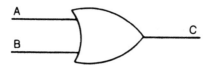

The OR Gate

In the last two chapters we have seen that there are basic digital gates that perform the Boolean algebra NOT and AND functions. Not surprisingly, the next primary type of digital gate performs the OR function. It is called an OR gate, and its schematic symbol is shown in Fig. 4-1. The truth table is given in Table 4-1.

As with the AND function, we can simulate the OR function with a simple mechanical switching circuit. This is illustrated in Fig. 4-2. Here the switches are in parallel. At least one of the switches must be at logic 1 (closed) for the output to be at logic 1 (lamp lit). The output will be a logic 0 if, and only if, all of the switches are open (logic 0). Of course, as many switches as desired may be added in parallel. As long as one or more switches are closed (logic 1), the lamp will light (logic 1 output).

A simple electronic OR gate made up of two diodes and a resistor is shown in Fig. 4-3. The diodes are forward biased and can conduct only if a high voltage (logic 1) is presented to the appropriate inputs. If both inputs are grounded (logic 0), there will be no signal at the output. In other words, the output will also be a logic 0. But if a voltage is applied to either or both of the diodes, current will flow and most of the input voltage will appear at the output, signifying a logic 1.

Figure 4-4 shows a transistorized version of the basic OR gate. Each transistor is either cutoff (logic 0), or saturated (logic 1), depending on the voltage fed to its base. A voltage (logic 1) will

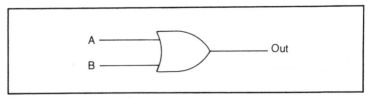

Fig. 4-1. This symbol is used in diagrams to represent an OR gate.

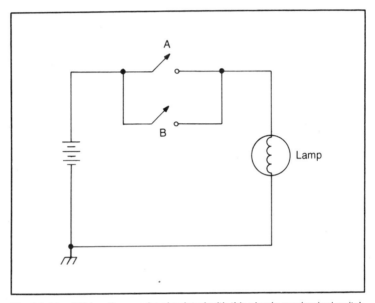

Fig. 4-2. The OR function can be simulated with this simple mechanical switch circuit.

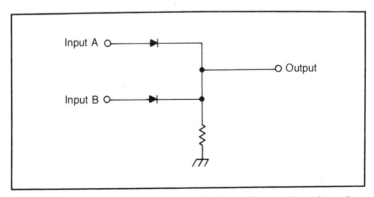

Fig. 4-3. The OR function can be performed with two diodes and a resistor when connected into this circuit.

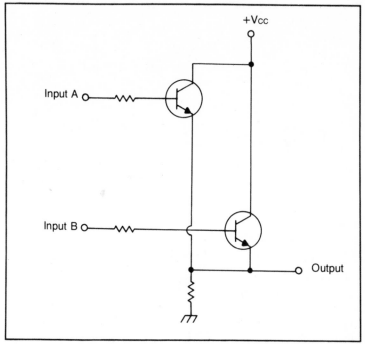

Fig. 4-4. An improved OR gate circuit can be built using transistors.

appear at the output only if at least one of the transistors are conducting. If both transistors are cut off (logic 0), there will be essentially no voltage at the output (logic 0).

OR gates in practical electronic digital circuits are generally in integrated circuit form. Readily-available ICs containing four two-input OR gates are widely used.

MULTIPLE-INPUT OR GATES

The OR function can also work with multiple inputs. Figure 4-5

Inputs		Output
A	B	
0	0	0
0	1	1
1	0	1
1	1	1

Table 4-1. The Truth Table for an OR Gate Is the Same as the One for the Boolean Algebra Function.

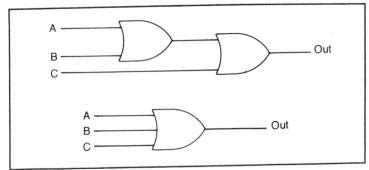

Fig. 4-5. OR gating can be done with more than two inputs.

shows a three-input OR gate made up of two standard two-input OR gates, and a single three-input OR gate. The truth table for a three-input OR gate is shown in Table 4-2. An OR gate may have as many inputs as required for any given application. The output will

Inputs			Output
A	B	C	
0	0	0	0
0	0	1	1
0	1	0	1
0	1	1	1
1	0	0	1
1	0	1	1
1	1	0	1
1	1	1	1

Table 4-2. The Truth Table for a Three-Input OR Gate Is Simply an Extension of the Two-Input Version.

be a logic 0, only if all of the inputs are at logic 1. One or more logic 1 inputs will produce a logic 1 output.

The OR function may be altered by combining OR gates with inverters. For example, inverting one of the inputs, as shown in

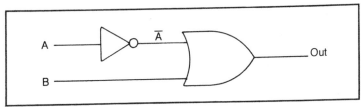

Fig. 4-6. A new gating function can be created by inverting one of the inputs to an OR gate.

Inputs			Output
A	B	$\overline{A}$	
0	0	1	1
0	1	1	1
1	0	0	0
1	1	0	1

Table 4-3. This Truth Table Shows the Results of Inverting One of the Inputs to an OR Gate.

Fig. 4-6, will produce a gate that works with the truth table shown in Table 4-3.

THE NOR GATE

More commonly used, is the inverted output OR gate, shown in Fig. 4-7. This will produce the exact opposite of the ordinary OR function, as indicated in the truth table of Table 4-4. This function is

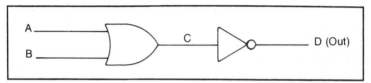

Fig. 4-7. An inverter at the output can reverse the action of an OR gate.

so widely used, it is available in IC form. It is called a NOT OR gate, or a NOR gate. The output is a logic 1 if, and only if, neither output A NOR output B is a logic 1. The NOR gate may be expanded for multiple inputs. The output will be a logic 0 whenever one or more of the inputs is at logic 1. All of the inputs must be at logic 0 for a logic 1 output. The standard symbol for a NOR gate is illustrated in Fig. 4-8.

THE EXCLUSIVE-OR GATE

There is another important variation on the basic OR gate.

Table 4-4. This Truth Table Shows How the NOR Function Is the Exact Opposite of an OR Gate.

Inputs		Output
A	B	
0	0	1
0	1	0
1	0	0
1	1	0

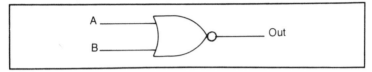

Fig. 4-8. An inverted output OR gate becomes a NOR gate.

Inputs A B	Output
0 0	0
0 1	1
1 0	1
1 1	0

**Table 4-5. This Truth Table
Shows How the Exclusive-OR Function
Differs from the Standard OR Function**

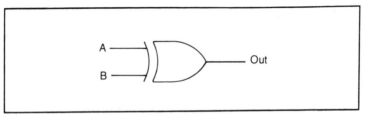

Fig. 4-9. A variation on the standard OR gate is the Exclusive-OR gate.

**Table 4-6. Here is the
Truth Table for an Exclusive-
OR Gate with an Inverted Output.**

Inputs		Output	
A	B	C	D
0	0	0	1
0	1	1	0
1	0	1	0
1	1	0	1

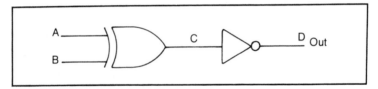

Fig. 4-10. Inverting the output of an Exclusive-OR gate produces a logic equality detector.

36

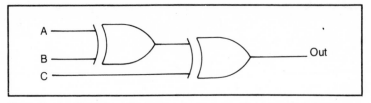

Fig. 4-11. The Exclusive-OR function can be extended for multiple inputs.

Table 4-7. The Truth Table for a Three-Input Exclusive-OR Gate is an Extension of the One for a Two-Input Exclusive-OR Gate.

Inputs			Output
A	B	C	
0	0	0	0
0	0	1	1
0	1	0	1
0	1	1	0
1	0	0	1
1	0	1	0
1	1	0	0
1	1	1	0

This is the Exclusive-OR gate, or, as it is sometimes called, the X-OR gate. In this case, the output will be a logic 1 if one, but not both, of the inputs is a logic 1. If both inputs are logic 0's, or if both are logic 1's, the output will be a logic 0.

Another way to look at the Exclusive-OR gate is that the output is a logic 1 if, and only if, the two inputs have different values. If the two input signals are the same, the output will be a logic 0. An Exclusive-OR gate could be called a "Logic Difference Detector." The schematic symbol for an Exclusive-OR gate is shown in Fig. 4-9, and the truth table is given in Table 4-6.

The action of an Exclusive-OR gate can be reversed by inverting the output, as illustrated in Fig. 4-10. In this case the output will be a logic 1 if both input values are the same (both 1's, or both 0's), but a logic 0, if they are different (one input at logic 0, and the other at logic 1). This operation is summarized in the truth table shown in Table 4-6.

While they are not commonly used, multiple input Exclusive-OR gates are feasible. A three-input Exclusive-OR gate is illustrated in Fig. 4-11, and its truth table is shown in Table 4-7. The output is a logic 1 if only one of the inputs is a logic 1, and the others are all logic 0's. If all the inputs are at logic 0, or two or more inputs are at logic 1, the output will be a logic 0.

Chapter 5

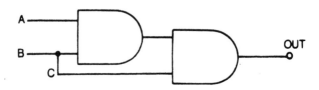

Combining Digital Gates

In practical digital electronic circuits, single gates are rarely encountered. More commonly a number of gate packages will be combined to create more complex gating systems. Virtually any logic gating pattern may be achieved by combining the basic types of gates described in the previous chapters.

CREATING OR GATES FROM AND GATES

Actually, while there are three basic gate types (AND, OR, and NOT), only two are needed to set up any gating pattern. An OR gate can be simulated with an AND gate and three inverters, as shown in Fig. 5-1. Table 5-1 shows the logic states throughout the circuit. Notice that this circuit performs the NOR operation at point E, so the output inverter may be left off if a NOR gate is required.

AND gates and inverters can also be combined to perform the Exclusive-OR operation. One method is illustrated in Fig. 5-2. A somewhat more elegant approach is illustrated in Fig. 5-3. This demonstrates an important point in digital design—there is usually more than one way to accomplish any specific gating task.

CREATING AND GATES FROM OR GATES

The reverse procedure may also be used. If no AND gate package is available, its function may be simulated using an OR gate and three inverters. The circuit is shown in Fig. 5-4, and the truth

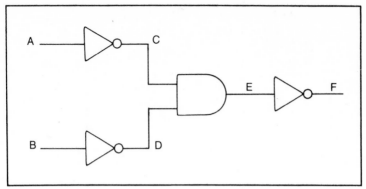

Fig. 5-1. The OR function can be performed with an AND gate and three inverters.

table is given in Table 5-2. Once again, the final inverter may be omitted if a NAND operation is required.

Do you see the similarity here to the gating circuit of Fig. 5-1? You certainly should. Inverting the inputs of a gate simulates the negated (inverted output) version of the opposite type of gate. Inverting the inputs of an AND gate results in a NOR operation. Inverting the inputs of an OR gate simulates the action of a NAND gate.

The same principle holds true with NAND and NOR gates. Inverting the inputs of a NAND gate simulates an OR gate, and a NOR gate with inverted inputs will behave like an AND gate.

TWO-INPUT COMBINATIONS

There are sixteen possible output patterns for a two-input logical gating circuit. These are summarized in Table 5-3. Some combinations are trivial; others are extremely useful.

Table 5-1. This Truth Table Demonstrates How the OR Function Can Be Simulated with an AND Gate and Three Inverters (See Fig. 5-1).

		Inputs		NOR Output	OR Output
A	B	C	D	E	F
0	0	1	1	1	0
0	1	1	0	0	1
1	0	0	1	0	1
1	1	0	0	0	1

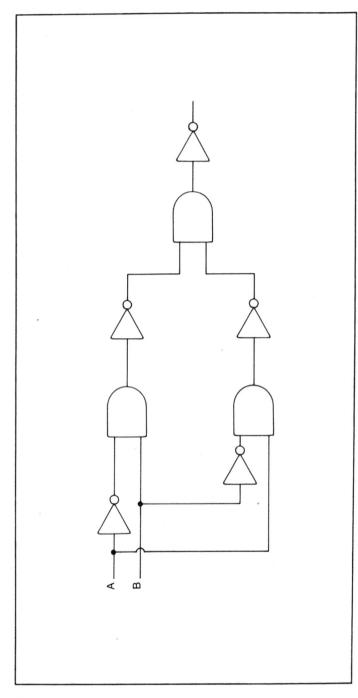

Fig. 5-2. The Exclusive-OR function can be performed by a combination of AND gates and inverters.

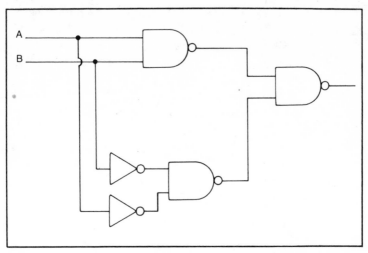

Fig. 5-3. Here is another way to perform the exclusive-OR function with a combination of AND gates and inverters.

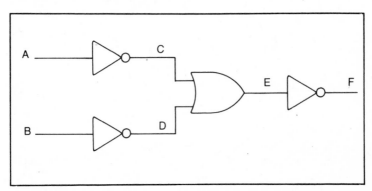

Fig. 5-4. The AND function can be performed with an OR gate and three inverters.

Table 5-2. This Truth Table Demonstrates How the AND Function Can Be Simulated with an OR Gate and Three Inverters (See Fig. 5-4).

	Inputs			NAND Output	AND Output
A	B	C	D	E	F
0	0	1	1	1	0
0	1	1	0	1	0
1	0	0	1	1	0
1	1	0	0	0	1

Table 5-3. Two-Input Logic Gates Can Produce Up to 16 Output Patterns.

Inputs		Outputs										
A	B	C	D	E	F	G	H	I	J	K	L	M
0	0	0	0	0	0	1	0	0	1	0	1	1
0	1	0	0	0	1	0	0	1	0	1	0	1
1	0	0	0	1	0	0	1	0	0	1	1	0
1	1	0	1	0	0	0	1	1	1	0	0	0

Inputs		Outputs				
A	B	N	O	P	Q	R
0	0	0	1	1	1	1
0	1	1	0	1	1	1
1	0	1	1	0	1	1
1	1	1	1	1	0	1

Output pattern C is obviously trivial and useless. The output is a logic 0, regardless of the signal at either of the inputs. Output pattern D should look familiar. The output is a logical 1 only when both inputs are logic 1's. Any other combination of inputs results in a logic 0 output. Of course, this is the basic AND function.

In output pattern E, the output is a logic 1 if, and only if, input A is a logic 1, and input B is a logic 0. All other input combinations will leave the output at logic zero. This gating action can be created with an AND gate and an inverter, as shown in Fig. 5-5. The B input is inverted so that the AND gate will see a 0 as a 1. The AND gate will produce a logic 1 output when A = 1, and $\overline{B}$ = 1 (i.e., B = 0).

The same pattern can also be created with a NOR gate and an inverter. This is illustrated in Fig. 5-6. Here, the A input is inverted, so a logic 1 at this input will appear as a 0 to the NOR gate. The output will be a 1 if, and only if, B = 0 and $\overline{A}$ = 0 (i.e., A = 1).

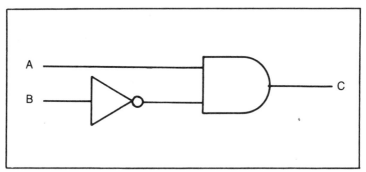

Fig. 5-5. This is one way to set up an "A but not B" gate.

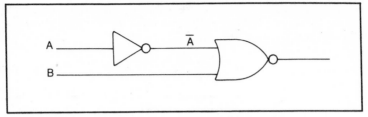

Fig. 5-6. This is an alternate way to set up an "A but not B" gate.

These two circuits perform in exactly the same way.

Output pattern F is similar to output pattern E, except the inputs are reversed. The output is a 1 if, and only if A = 0, and B = 1. This pattern can be generated with the circuits shown in Figs. 5-5 and 5-6 if the inverter in each case is moved to the opposite input.

Moving on to output pattern G we have another familiar combination. The output is a 1 if, and only if neither input A nor input B is a logic 1 (both inputs are at logic 0). This is the standard NOR function.

Output pattern H is another trivial combination. The output is logic 1 if input A is logic 1, and a logic 0 if input A is at logic 0. The state of input B doesn't matter. Of course, to achieve this pattern you only need signal A, and no gate at all. Similarly, output pattern I merely duplicates the value of input B, while ignoring input A altogether.

Output pattern J has an output of 1 whenever the two inputs are equal (both 0's, or both 1's). If the inputs have opposite values, the output will be a logic 0. We discussed this pattern in Chapter 4. It is the mirror image of the Exclusive-OR function, so it can easily be achieved by inverting the output of an Exclusive-OR gate.

Output pattern K is the opposite of pattern J. The output is a 1 only if the two inputs have differing values. This, of course, is the straight Exclusive-OR function.

Output pattern L consists of a logic 1 when input B is a logic 0, and a logic 0 when input B is at logic 1. The value of input A is irrelevant. This pattern can be achieved simply by inverting B. By

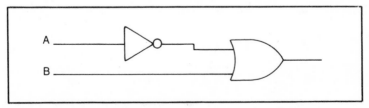

Fig. 5-7. Here is a circuit for obtaining logic pattern 0 from Table 5-3.

43

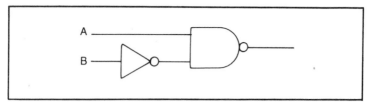

Fig. 5-8. There is more than one way to generate logic pattern 0 from Table 5-3.

the same token, output pattern M is nothing more than the inversion of input A. Input B is ignored.

The next output pattern is N, and it is another familiar one. The output is a logic 0 only when both inputs are 0's. If either or both of the inputs goes to logic 1, the output will also be a logic 1. This is the standard OR function.

In output pattern 0, the output is a logic 1 unless input A is at 0 *and* input B is at 1, in which case the output is logic 0. This pattern

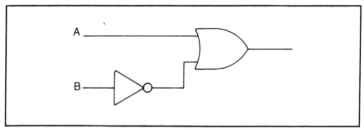

Fig. 5-9. Here is the first method for generating logic pattern P from Table 5-3.

may be achieved with an OR gate, as shown in Fig. 5-7, or with a NAND gate, as illustrated in Fig. 5-8.

Output pattern P is very similar to pattern 0. The only difference is that the inputs are reversed. This time input A must be a 1, and input B must be a 0 for a 0 to appear at the output. All other input combinations will result in a logic 1 at the input. Figures 5-9, and 5-10 illustrate the two simplest ways to produce this output pattern.

Another familiar pattern is output pattern Q. The output is a logic 0 if, and only if, both inputs are at logic 1. If one or both of the

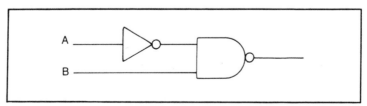

Fig. 5-10. This is a second method for obtaining logic pattern P from Table 5-3.

inputs is a 0, the output will be a 1. This, of course, is the opposite of the basic AND operation. In other words, this pattern can be produced with a simple NAND gate.

The last possible output pattern, R, is completely trivial and useless. It is always at logic 1, regardless of the value of either input.

As this brief summary shows, any output pattern for two logical inputs can easily be achieved by mixing just a few of the basic types of gates in simple combinations.

MULTIPLE-INPUT GATES

Logic circuits may certainly use more than two inputs, and usually do. This can complicate matters. For three-input gates, the

Inputs					Output
A	B	C	D	E	
0	0	0	0	0	0
0	0	0	0	1	0
0	0	0	1	0	1
0	0	0	1	1	1
0	0	1	0	0	0
0	0	1	0	1	0
0	0	1	1	0	1
0	0	1	1	1	1
0	1	0	0	0	1
0	1	0	0	1	1
0	1	0	1	0	1
0	1	0	1	1	1
0	1	1	0	0	1
0	1	1	0	1	1
0	1	1	1	0	1
0	1	1	1	1	1
1	0	0	0	0	0
1	0	0	0	1	0
1	0	0	1	0	1
1	0	0	1	1	1
1	0	1	0	0	0
1	0	1	0	1	0
1	0	1	1	0	1
1	0	1	1	1	1
1	1	0	0	0	1
1	1	0	0	1	1
1	1	0	1	0	1
1	1	0	1	1	1
1	1	1	0	0	1
1	1	1	0	1	1
1	1	1	1	0	1
1	1	1	1	1	1

Table 5-4. This Is a Simple Truth Table for a Five-Input Gating Circuit.

Inputs	Output
A B C D E	
x 0 x 0 x	0
x 0 x 1 x	1
x 1 x 0 x	1
x 1 x 1 x	1
(x = "don't care")	

Table 5-5. The Truth Table of Table 5-4 Can Be Simplified as Shown Here.

number of possible output combinations is increased to 64. For four-input circuits, this jumps to 256. A five-input logic gate can have any of 1024 possible output combinations. There is no theoretical limit to the number of inputs.

Obviously, we can not cover every possible combination in this book. But, by fully understanding the principles involved, you should be able to construct any gate your digital circuit requires.

Of course, in the vast number of combinations mentioned above, many are trivial. One or more of the inputs may be irrelevant to the output, allowing a simpler gating combination to be used. For example, the complex looking five-input truth table shown in Table 5-4 can be simplified to a standard OR gate for inputs B and D (Table 5-5). Inputs A, C, and E are irrelevant.

Some gate circuits also have multiple outputs in addition to multiple inputs. For example, a truth table for a typical three-input/two-output gating circuit is given in Table 5-6. Figure 5-11 shows a circuit that can be used to generate this logic pattern.

Usually, when designing a logic gating system, it is a good idea to draw up a truth table that accounts for all possible input combinations. An example is shown in Table 5-7.

Except for relatively simple truth tables, it is generally difficult to work from a complete truth table directly. Fortunately,

Table 5-6. This Is the Truth Table for the Three-Input/Two-Output Example Described in the Text.

Inputs			Outputs	
A	B	C	D	E
0	0	0	0	1
0	0	1	0	1
0	1	0	0	1
0	1	1	0	0
1	0	0	0	1
1	0	1	1	0
1	1	0	0	1
1	1	1	1	0

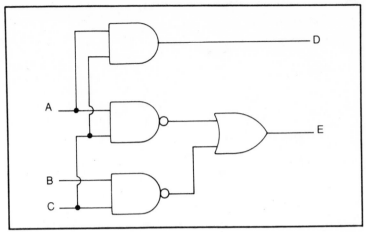

Fig. 5-11. Digital gate circuits may have multiple inputs and outputs, as in this typical three-input/two-output gate circuit.

most truth tables can be simplified. One or more input may be irrelevant under certain circumstances. For instance, in our example, the output is always a logic 1 when inputs A and D are both at logic 0. In this case, the values of B and C are irrelevant. The circuit does not care about B and C when A = 0 and D = 0. We can simplify the truth table by reducing four input combinations (0000, 0010, 0100, and 0110) with one (0xx0), where the letter "x" is used to represent a "don't care" condition. The output will be the same if the don't care input is 0 or 1.

Inputs				Output
A	B	C	D	
0	0	0	0	1
0	0	0	1	0
0	0	1	0	1
0	0	1	1	1
0	1	0	0	0
0	1	0	1	0
0	1	1	0	0
0	1	1	1	1
1	0	0	0	0
1	0	0	1	0
1	0	1	0	0
1	0	1	1	1
1	1	0	0	1
1	1	0	1	1
1	1	1	0	1
1	1	1	1	1

Table 5-7. The Design Example Described in the Text Uses This Truth Table.

47

Inputs				Output
A	B	C	D	
0	x	x	0	1
0	0	0	1	0
0	0	1	1	1
0	1	0	1	0
0	1	1	1	1
1	0	0	0	0
1	0	0	1	0
1	0	1	0	0
1	0	1	1	1
1	1	x	x	1

Table 5-8 The Truth Table of Table 5-7 Can Be Revised as Shown Here.

Table 5-8 is a simplified version of Table 5-7. Notice how much easier it is to take in all the information, now that the irrelevant combinations have been eliminated.

The next step is to examine the truth table closely for patterns. In this example, the output is a logic 1 if A and D = 0, or C and D = 1 or A and B = 1. All other combinations produce a logic 0 at the output. We can easily reduce this to Boolean algebra form. The gate will perform this complex equation—(NOT (A AND D)) OR (C AND D) OR (A AND B). In other words, we can create this logic pattern with two AND gates, a NAND gate, and a three-input (or two two-input) OR gate. The circuit is shown in Fig. 5-12.

Many practical gating circuits will be more complicated than

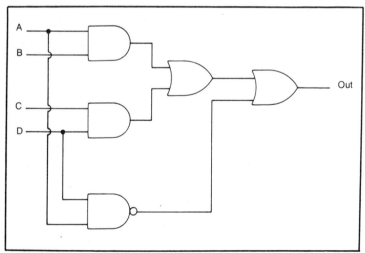

Fig. 5-12. This circuit will generate the logic pattern for the first design example as outlined in Table 5-7.

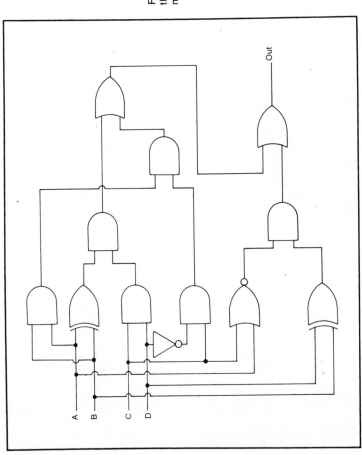

Fig. 5-13. This circuit will produce the complex logic pattern summarized in Table 5-9.

Inputs				Output
A	B	C	D	
0	0	0	0	0
0	0	0	1	1
0	0	1	0	0
0	0	1	1	0
0	1	0	0	1
0	1	0	1	0
0	1	1	0	0
0	1	1	1	1
1	0	0	0	0
1	0	0	1	0
1	0	1	0	0
1	0	1	1	1
1	1	0	0	0
1	1	0	1	0
1	1	1	0	1
1	1	1	1	0

**Table 5-9. Here Is a
Truth Table for a Somewhat
More Complex Gating Circuit.**

this example. For instance, consider the truth table and circuit in
Table 5-9 and Fig. 5-13. But any design problem in a logic system
can be solved in the same manner. Break the required truth table
down, eliminate irrelevant combinations ("don't care" inputs), and
seek out output patterns that can be performed with basic gates.

TYPICAL APPLICATIONS FOR DIGITAL GATES

There are thousands of potential applications for digital gates.

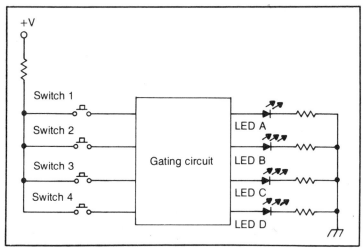

Fig. 5-14. Here is a block diagram for the second design example.

50

These applications range from simple state indicators through giant computers. There is certainly no way to cover all of the possible applications in a single book.

Control functions and automatic switching are "naturals" for digital gates. Each switch on a control panel could control a specific function while certain combinations of simultaneously pressed buttons could perform different functions. To illustrate this concept, we will now work through an example.

A block diagram for this example problem is shown in Fig. 5-14. Four push-button switches (numbered one through four) feed logic 1's into the gating circuit when closed. The LEDs will light according to what switches are closed. In a practical circuit, the LEDs would probably be replaced with control circuitry. For instance, a logic controlled tape recorder might have the buttons labeled PLAY, RECORD, REWIND, and STOP, with the gating circuit controlling the appropriate motors and amplifier circuits.

In this example, we'll assume we want the LEDs to indicate what numerical value is being entered. LED A would be lit only when switch 1, and no other switch, is depressed. Similarly, LED B will light if, and only if, switch 2 is the only closed switch. LED C can be lit by depressing only switch 3, or by simultaneously closing both switches 1 and 2 (1 + 2 = 3), while leaving switches 3 and 4 open. Finally, LED 4 can be turned on by closing switch 4, and no other switch, or by closing switches 1 and 3 together, while switches 2 and 4 are kept open. Any other combination of depressed

Table 5-10. This Is the Truth Table for the Second Design Example.

1	2	3	4	A	B	C	D
0	0	0	0	0	0	0	0
0	0	0	1	0	0	0	1
0	0	1	0	0	0	1	0
0	0	1	1	0	0	0	0
0	1	0	0	0	1	0	0
0	1	0	1	0	0	0	0
0	1	1	1	0	0	0	0
1	0	0	0	1	0	0	0
1	0	0	1	0	0	0	0
1	0	1	0	0	0	0	1
1	0	1	1	0	0	0	0
1	1	0	0	0	0	1	0
1	1	0	1	0	0	0	0
1	1	1	0	0	0	0	0
1	1	1	1	0	0	0	0

switches will leave all four LEDs dark.

The desired action of this sample circuit is summarized in Table 5-10. Notice that there are four inputs (1, 2, 3, and 4) and four outputs (A, B, C, and D). There is little to be gained by over-complicating things and trying to solve the problem all at once. We'll make the task easier by breaking things down.

First, examine the output patterns. Are there any similarities? For instance, is one an inversion of another, or can two be ORed and ANDed together to produce a third?

In this particular example, the answers to these questions are all "no." The four output patterns are distinct and separate. In this case, the best approach would be to deal with each output separately, at least, for the time being.

Table 5-11 is a simplification of the main truth table. Outputs B, C, and D are temporarily eliminated, so we can concentrate on the gating required to generate output pattern A.

Examine this partial truth table to see if any of the inputs are irrelevant (that is, their values may be replaced by X, or "don't care"). In our example here, only a single input combination will produce a logic 1 output, so obviously all of the inputs are relevant, and must be dealt with.

The output A will be a logic 1, if, and only if, input 1 is at logic 1, and the other three inputs are at logic 0. This is somewhat similar to the "A but NOT B" pattern discussed earlier. As you should recall, that output pattern can be achieved by inverting one of the inputs to

Inputs				Output
1	2	3	4	A
0	0	0	0	0
0	0	0	1	0
0	0	1	0	0
0	0	1	1	0
0	1	0	0	0
0	1	0	1	0
0	1	1	0	0
0	1	1	1	0
1	0	0	0	1
1	0	0	1	0
1	0	1	0	0
1	0	1	1	0
1	1	0	0	0
1	1	0	1	0
1	1	1	0	0
1	1	1	1	0

Table 5-11. The Truth Table of Table 5-10 Can Be Simplified As Shown Here to Solve for Output A.

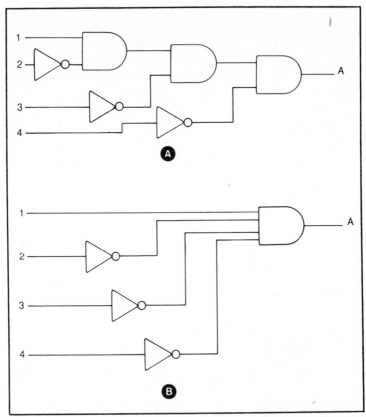

Fig. 5-15. Here is one way to generate output A from the second design example.

an AND gate. By extension, we can produce the desired output pattern by inverting three of the inputs of a four-input AND gate, as illustrated in Fig. 5-15. This is certainly the most straightforward method of producing output pattern A. But is it the best and most economical method possible?

Let's try an alternative approach. Input 1 must be a logic 1, while neither input 2 NOR input 3 NOR input 4 may be a 1. By following this logic, we can set up the gating circuit illustrated in Fig. 5-16.

We will follow the logic through this circuit. First, inputs 2 and 3 are ORed. This means a is a logic 1 if either input 2 OR input 3 is a 1. Only if both of these inputs is 0, will a be at logic 0. In other words a = 2 OR 3, or a = 2 + 3.

Sub-output a is then NORed with input 4. The output of this gate (b) will be a 1 if, and only if, both a (i.e., both 2 and 3) and 4 are

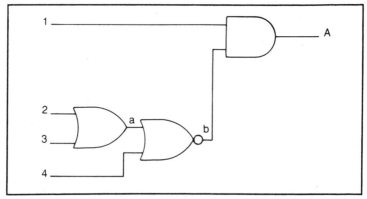

Fig. 5-16. There is more than one way to produce output A from the second design example.

at logic 0. If either input 4 or a (input 2 or input 3 or both) is at logic 1, then sub-output b will be 0.

Finally, sub-output b is ANDed with input 1. Both b and 1 must be at logic 1 for a logic 1 at the final output A. This will only occur when input 1 is at logic 1, while a (both input 1 and input 3) and input 4 are at logic 0. All other combinations will produce logic 0 at output A. That is exactly the output pattern we want.

Now we will move on to output B. The partial truth table for this output is given in Table 5-12. As with output A, output B will be a logic 1 only for a single combination of inputs. Output A was high when input 1 and no other input was at logic 1. Similarly, for output

	Inputs			Output
1	2	3	4	B
0	0	0	0	0
0	0	0	1	0
0	0	1	0	0
0	0	1	1	0
0	1	0	0	1
0	1	0	1	0
0	1	1	0	0
0	1	1	1	0
1	0	0	0	0
1	0	0	1	0
1	0	1	0	0
1	0	1	1	0
1	1	0	0	0
1	1	0	1	0
1	1	1	0	0
1	1	1	1	0

Table 5-12. In the Second Step of the Design Example, We Concentrate on the Truth Table for Output B.

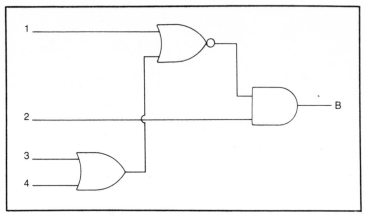

Fig. 5-17. The circuit for generating output B for the second design example is similar to the one for output A.

B, input 2 must be logic 1, while the other three inputs are at logic 0 for the output to be a 1. Any other combination of inputs will produce logic 0 at the output. Output B is so similar to output A, that we can use the same circuit and simply redirect the inputs. This is shown in Fig. 5-17.

Output C (whose truth table is shown in Table 5-13), is at logic 1 for two of the possible input combinations. When input 3 is at logic 1, and the other three inputs are at logic 0, output C will go to logic 1. This is similar to the patterns we've already covered for outputs A and B, and can be achieved with the same type of circuit.

Table 5-13. Next We Rearrange the Truth Table from Table 5-10 to Solve for Output C.

\multicolumn{4}{c}{Inputs}				Output
1	2	3	4	C
0	0	0	0	0
0	0	0	1	0
0	0	1	0	1
0	0	1	1	0
0	1	0	0	0
0	1	0	1	0
0	1	1	0	0
0	1	1	1	0
1	0	0	0	0
1	0	0	1	0
1	0	1	0	0
1	0	1	1	0
1	1	0	0	1
1	1	0	1	0
1	1	1	0	0
1	1	1	1	0

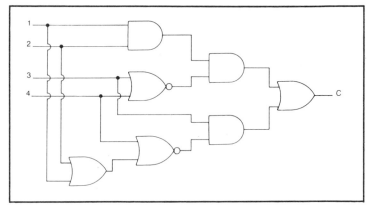

Fig. 5-18. The circuit for output C will produce a logic 1 output for two input combinations.

But output C will also be a logic 1 when both input 1 and input 2 are at logic 1 (since $1 + 2 = 3$), and input 3 and input 4 at logic 0. Inputs 1 and 2 can be ANDed, and inputs 3 and 4 can be NORed, then the outputs can be ANDed to produce a logic 1 output for this particular input combination. This output can then be ORed with the output of the circuit that produces a logic 1 output when input 3, and no other input is a 1. The complete gating circuit for output C is shown in Fig. 5-18.

Finally, the simplified truth table for output D is given in Table 5-14. This pattern is similar to output C—the output is a logic 1

Inputs				Output
1	2	3	4	D
0	0	0	0	0
0	0	0	1	1
0	0	1	0	0
0	0	1	1	0
0	1	0	0	0
0	1	0	1	0
0	1	1	0	0
0	1	1	1	0
1	0	0	0	0
1	0	0	1	0
1	0	1	0	1
1	0	1	1	0
1	1	0	0	0
1	1	0	1	0
1	1	1	0	0
1	1	1	1	0

Table 5-14. Finally,
We Simplify the Truth Table
of Table 5-10 to Solve for Output D.

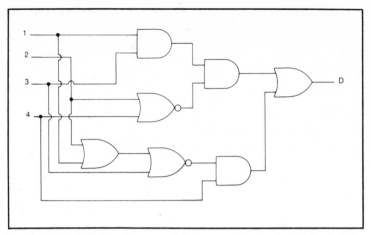

Fig. 5-19. Output D from the second design example can be obtained with a circuit similar to the one for output C.

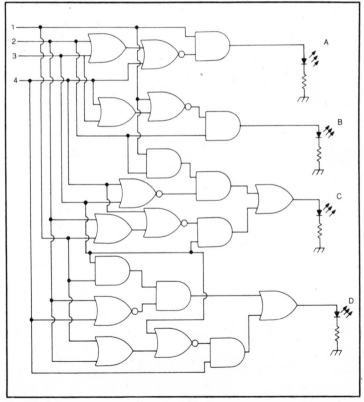

Fig. 5-20. This circuit is the complete solution for the second design example.

AND Gates	OR Gates	NOR Gates
8	6	6
(2 quad packages)	(1½ quad packages)	(1½ quad packages)

Total Gates Used = 19
Total ICs Required = 6

when input 4, and no other input is a logic 1, or when inputs 1 and 3 are at logic 1 and inputs 2 and 4 are at logic 0. Obviously, the gating circuit for output D can be similar to that for output C, with the inputs rerouted as necessary. The gating circuit for output D is illustrated in Fig. 5-19.

Now that we have solved each of the outputs, we can put everything back together to come up with the complete circuit shown in Fig. 5-20.

Is this the best solution? Can you produce the same output patterns with fewer gates? The gates required for this solution are summarized in Table 5-15. You may want to try to come up with a more efficient solution on your own as an exercise.

Table 5-16 shows the truth table for yet another design example. Once again, we have four inputs (A, B, C, and D) and four

**Table 5-16. Here Is the Truth Table for the
Third Design Example, as Discussed in the Text.**

Inputs				Outputs			
A	B	C	D	E	F	G	H
0	0	0	0	0	0	1	0
0	0	0	1	0	0	1	0
0	0	1	0	1	0	0	0
0	0	1	1	1	1	0	0
0	1	0	0	1	0	0	0
0	1	0	1	1	0	0	0
0	1	1	0	0	0	1	0
0	1	1	1	0	1	1	1
1	0	0	0	1	0	0	0
1	0	0	1	1	0	0	0
1	0	1	0	0	0	1	0
1	0	1	1	0	1	1	1
1	1	0	0	0	1	1	1
1	1	0	1	0	1	1	1
1	1	1	0	1	1	0	0
1	1	1	1	1	1	0	0

Inputs				Output
A	B	C	D	E
0	0	0	x	0
0	0	1	x	1
0	1	0	x	1
0	1	1	x	0
1	0	0	x	1
1	0	1	x	0
1	1	0	x	0
1	1	1	x	1

Table 5-17. The Truth Table from Table 5-16 Can Be Simplified Like This to Solve for Output E.

outputs (E, F, G, and H). We can again solve each output independently, but if you examine the truth table carefully, you should be able to discover a few short cuts.

For one thing, output G is always at logic 0 when output E is at logic 1, and vice versa. Rather than setting up a complete gating circuit to generate output G, we can simply invert output E.

Now examine output H in relation to the other outputs. Notice how this output is a 1 only when both output F and output G are at logic 1. In other words, output F and output G can be ANDed to produce output H. Since outputs G and H can be obtained from the other outputs, we only have to design gates for output E and output F.

Table 5-17 is a reduced truth table for output E. Notice that the column for input D is a string of x's, or "don't care's." The status of input D is irrelevant to output E. Only the first three inputs (A, B, and C) determine the output pattern at E.

Four of the eight possible input combinations produce a logic 1 output. They are 001, 010, 100, and 111. What similarities can we find between these input combinations? There are several potential solutions, of course, but I think this is the simplest. When input A is

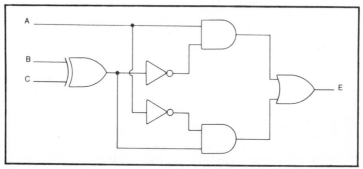

Fig. 5-21. Output E in the third design example can be generated with this circuit.

Inputs				Output
A	B	C	D	F
0	0	0	0	0
0	0	0	1	0
0	0	1	0	0
0	0	1	1	1
0	1	0	0	0
0	1	0	1	0
0	1	1	0	0
0	1	1	1	1
1	0	0	0	0
1	0	0	1	0
1	0	1	0	0
1	0	1	1	1
1	1	0	0	1
1	1	0	1	1
1	1	1	0	1
1	1	1	1	1

Table 5-18. The Truth Table for Output F in the Third Design Example Is Shown Here.

a logic 0, inputs B and C have unequal values, and when input A is a logic 1, input B must equal input C. This suggests inputs B and C can be combined with an Exclusive-OR gate, or its inversion. This sub-output is then ANDed with either A or $\overline{A}$ (NOT A), as appropriate. A circuit for generating this output pattern is illustrated in Fig. 5-21.

Next, we must solve for output F. The reduced truth table for this output is shown in Table 5-18.

All four of the inputs are of significance for this output, so there are 16 possible input combinations, of which, seven will produce a logic 1 at the output. These input combinations are 0011, 0111, 1011, 1100, 1101, 1110, and 1111. What are the similarities between these output combinations?

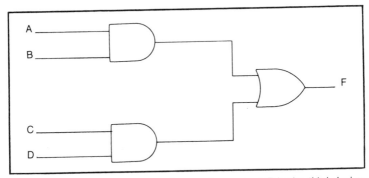

Fig. 5-22. This circuit can be used to produce output F in the third design example.

60

Table 5-19. Output G in the Third Example Is Simply the Inversion of Output E.

Output E	Output G
0	1
1	0

In all of these input combinations, at least two of the inputs are at logic 1. A little closer examination reveals that either inputs A and B are both logic 1, or inputs C and D are at logic 1, or all four inputs are at logic 1. This implies the use of AND gates, with their outputs ORed together. The circuit is shown in Fig. 5-22.

Next, we come to output G. As the simplified truth table in Table 5-19 reminds us, this output is simply an inversion of output E.

Table 5-20. Output H in the Third Design Example Uses Output E and Output F as Inputs.

Output F	Output G	Output H
0	0	0
0	1	0
1	0	0
1	1	1

Table 5-20 shows the simplified truth table for output H. As mentioned earlier, output H can be obtained simply by ANDing output F and output G.

The Boolean algebra equations for each of the outputs are presented in Table 5-21. The complete gating circuit is illustrated in Fig. 5-23, and the required gates are summarized in Table 5-22. Can you design a simpler circuit (requiring fewer gates) than this one?

Look back at the complete truth table for this example, which

Table 5-21. The Solution for the Third Design Example Can Be Summarized with Boolean Algebra Equations.

$E = ((B \pm C) \cdot \overline{A}) + (A \cdot (\overline{B \pm C}))$	*((B Exclusive-OR C) AND NOT A) OR (A AND NOT (B Exclusive-OR C))
$F = (A \cdot B) + (C \cdot D)$	= (A AND B) OR (C AND D)
$G = \overline{E}$	= NOT E
$H = F \cdot G$	= F AND G

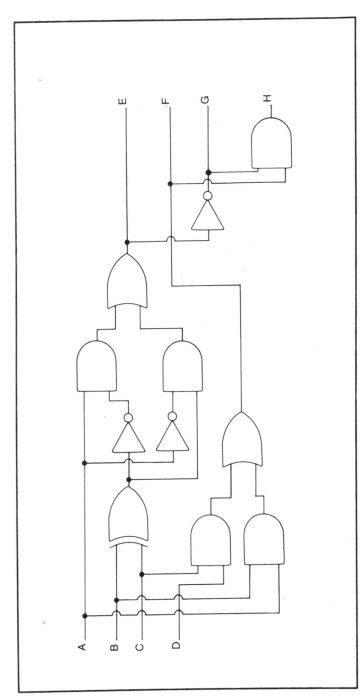

Fig. 5-23. Here is the complete solution circuit for the third design example.

**Table 5-22. Here Is a Summary of the Gates Required
for the Solution of the Third Design Example, as Shown in Fig. 5-23.**

AND gates	OR gates	EXCLUSIVE-OR gates	inverters
5	2	1	3
(packages required)			
1 ¼	½	¼	½
Gates required 11			
IC packages required 5			

was shown in Table 5-16. This truth table looks quite complex, but when we broke it down and worked step-by-step, the solution proved to be rather easy.

Don't let seemingly complicated logic tables throw you—most gating circuits are fairly simple once you've broken them down. Any gating circuit can be broken down to a combination of the three basic gate types (AND, OR, or NOT).

By the same token, virtually any digital circuit can be built from the basic gates, although many complex devices would probably require tens, hundreds, or even thousands of simple gates. Many applications (or subapplications) are so commonly used, they are available in direct IC form. In the next few chapters we will examine a number of these more complex digital devices.

Chapter 6

Multivibrators

Another important type of digital circuit is the multivibrator. As with the digital gate, a multivibrator's output is either a high voltage (logic 1) or a low voltage (logic 0) with no intermediate levels. There are three basic types of multivibrators. They are the monostable, the bistable, and the astable. We will discuss each of these and their applications in this and the following chapter.

MONOSTABLE MULTIVIBRATORS

The monostable multivibrator is primarily a timing device. By definition, this circuit has one stable output condition. That is the output is normally at logic 0, or at logic 1, depending on the specific design. When a brief trigger pulse is received by the input of the monostable multivibrator, the output switches to the opposite (unstable) logic state—i.e., from 0 to 1, or from 1 to 0. The output remains at this new logic state for a specific length of time (usually determined by a resistor/capacitor combination), then reverts back to the original, stable output condition until a new trigger pulse is received. This action is illustrated in Fig. 6-1. Because the monostable multivibrator outputs one pulse for each trigger signal received, it is often called a "one-shot."

As you might suspect, most of the common applications of the monostable multivibrator are related to timing. Because the length of the output pulse is predetermined and constant, trigger pulses of various lengths can be fed into a monostable circuit to be converted

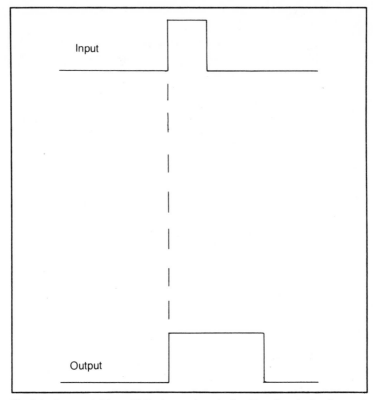

Fig. 6-1. A monostable multivibrator produces a fixed length output pulse each time it is triggered.

into a string of equal pulses, regardless of the length of the original pulses.

The output pulse from a monostable multivibrator is generally longer than the original trigger pulse. This means that, in effect, a monostable multivibrator stretches the input pulse. Such a circuit is occasionally referred to as a pulse stretcher. The original trigger pulse may be too short to properly trigger some other circuit, so a monostable multivibrator can be used to "stretch" the trigger pulse to the required length.

The pulse stretching function often comes in handy when indicators such as LEDs are used in digital circuits. Many of the signals in a digital circuit are just a tiny fraction of a second long. If a LED indicator is directly driven by such a signal, it will flash on and off so quickly, the human eye will be completely incapable of noticing the flash. A monostable multivibrator can be used to stretch the pulse driving the LED long enough to make the indication visible.

Yet another common application for monostable multivibrators in digital circuits is for signal clean-up. Many high speed digital circuits with manual entry switches include monostable circuits known as switch debouncers.

No mechanical switch is perfect, of course. When a switch is opened (or closed), it tends to bounce—that is, the contacts tend to rapidly touch each other and fly apart several times, before they settle into their new position. A high speed digital circuit can interpret each bounce as a new input signal, resulting in a count of ten or twenty, when only one was intended.

If the switch is used to trigger a monostable multivibrator, the output pulse can be long enough so that the switch contacts stop bouncing before it ends. Therefore, the driven circuit sees one and only one clean input pulse. The input and output signals of a typical switch debouncer are illustrated in Fig. 6-2.

To be used as a switch debouncer, the monostable multivibrator circuit must be of the nonretriggerable type. This means that once the circuit is triggered, additional trigger signals will be

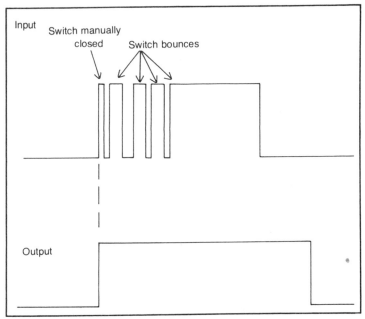

Fig. 6-2. A monostable multivibrator can be used to eliminate mechanical switch bouncing problems.

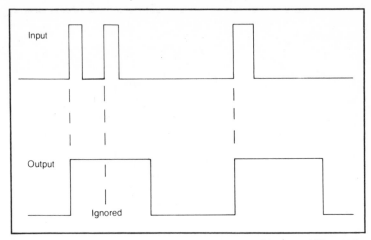

Fig. 6-3. A nonretriggerable monostable multivibrator will ignore additional trigger signals during its output cycle.

ignored until the output pulse is completed. See Fig. 6-3. Most practical circuits have a definite recovery time between when one output pulse is completed and the next can be begun. Some monostable multivibrator circuits, however, are retriggerable. This means, a new output cycle can be initiated before the previous cycle has been completed. This is illustrated in Fig. 6-4. Obviously the

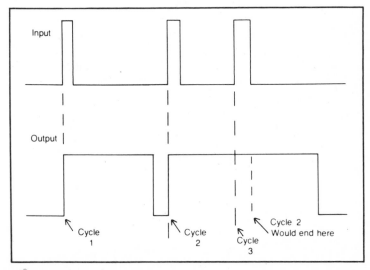

Fig. 6-4. A retriggerable multivibrator can start a new output cycle before the previous one is completed.

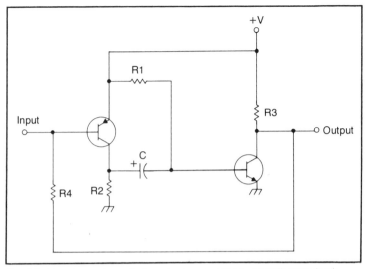

Fig. 6-5. Here is a schematic for a basic monostable multivibrator circuit.

choice between retriggerable, or nonretriggerable monostable multivibrators will depend on the specific application.

A basic schematic for a simple transistor based monostable multivibrator is shown in Fig. 6-5. Ordinarily both transistors are saturated, or conducting at their maximum level. If a brief positive pulse is fed to the input, both transistors will be cut off for a time determined by resistors R1 and R2, and the capacitor. While the transistors are cut off, the output will go high.

For some applications, monostable multivibrators can be built from digital inverters (see Fig. 6-6) or AND gates (see Fig. 6-7). Both of these circuits are triggered when the input switches from logic 0 (low) to logic 1 (high). Unlike most monostable circuits, the input pulse must be longer than the output pulse for these circuits to function. The length of the output pulse is determined by the resistor and capacitor values.

Dedicated monostable multivibrators in integrated circuit form

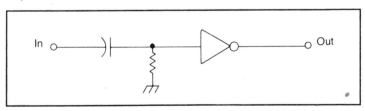

Fig. 6-6. A monostable multivibrator may be designed around a digital inverter.

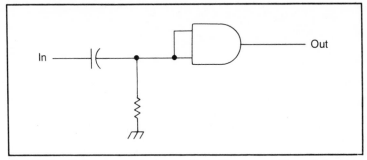

Fig. 6-7. The AND gate can also be used to build a monostable multivibrator.

are available. In the TTL family (see Chapter 12) we have the 74LS122, the 74123, the 74L123, and the 74LS123. Many designers, however, prefer to use the 555 timer IC for monostable functions. This is a linear device, but it is compatible with the major logic families. The pinout for the 555 is shown in Fig. 6-8. A basic monostable multivibrator circuit based on this device is shown in Fig. 6-9. The digital inverter is used to allow this circuit to work on positive pulses (normally at logic 0, with brief pulses to logic 1). If you are working with negative pulses (normally at logic 1, with brief pulses to logic 0), the inverter can simply be eliminated. The length of the output pulse is determined by resistor R1 and capacitor C1. The formula is:

$$T = 1.1 \ R1C1$$

It is a good idea to keep the value of R1 between 1 k and 10 megohms. $0.01 \ \mu F$ is a good standard value for capacitor C2. When more than one monostable is required within a single circuit, the 556 dual timer or the 558 quad timer may be used in place of individual 555 ICs.

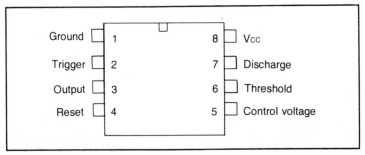

Fig. 6-8. The 555 is a Linear IC, but is widely used in digital circuits too.

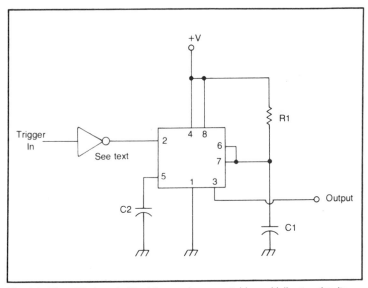

Fig. 6-9. The 555 timer is ideal for use in monostable multivibrator circuits.

BISTABLE MULTIVIBRATORS

A bistable multivibrator has two stable states. Each time this circuit is triggered, it reverses its output stage. Either output can be held indefinitely (as long as power is applied to the circuit, of course). In a sense, a bistable multivibrator "remembers" its last output state. The action of a bistable multivibrator is illustrated in Fig. 6-10.

Because of its ability to hold on to an output, a bistable multivibrator is often called latch. Another popular name for this same type of circuit is "flip-flop," because of the way the output can be switched between logic states. Many bistable multivibrators, or flip-flops, have two outputs, which are usually labeled Q and $\overline{Q}$ (NOT Q). Of course, $\overline{Q}$ is always the exact opposite of Q. That is, when Q = 1 then $\overline{Q}$ = 0, and vice versa.

There are a number of different types of flip-flops. This circuit can be put to work in countless applications. In fact, the flip-flop is so important in digital electronics that it deserves a chapter of its own. Bistable multivibrators will be discussed in depth in Chapter 7.

ASTABLE MULTIVIBRATORS

The third and final basic type of multivibrator is the astable. Where the monostable has one stable state, and the bistable has

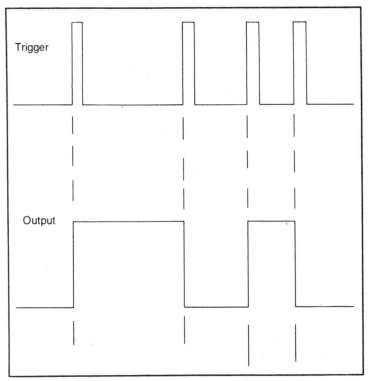

Fig. 6-10. A bistable multivibrator reverses its output state each time it is triggered.

two, the astable multivibrator has *no* stable output states. The output continuously switches back and forth between logic 0 and logic 1. In other words, an astable multivibrator is a form of oscillator, generating a signal known as a square wave. The output signal of a typical astable multivibrator is shown in Fig. 6-11. Generally, astable multivibrators have no input. They are self-running.

Figure 6-12 shows the circuit for a simple astable multivibrator. When power is first applied to this circuit, one of the transistors will start to conduct slightly faster than the other one. It

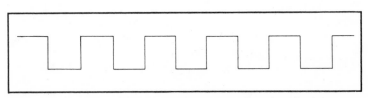

Fig. 6-11. An astable multivibrator generates a regular string of pulses.

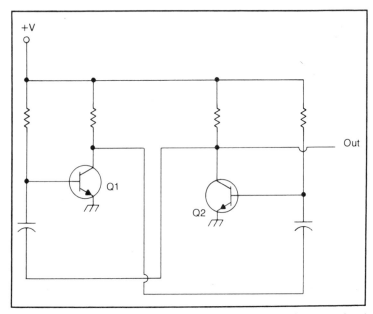

Fig. 6-12. In an astable multivibrator two transistors are alternately saturated and cut off.

doesn't matter which one this is. For the sake of discussion, we will assume it is transistor Q1.

Transistor Q1, once it begins to conduct, will rapidly reach its saturation point (i.e., its highest possible level of conduction). This will cause transistor Q2 to be cut off. This state of affairs will cause transistor Q2 to be cut off. This state of affairs will continue for a period determined by the time constant of the RC (resistor/capacitor) combination connected to Q1. The capacitor will then start to discharge through the other transistor, causing it to conduct. Transistor Q2 is shown saturated, and transistor Q1 is now cut off. The entire process is repeated. The result of all this is that the transistors are alternately switched on and off, causing the output to switch between a low and a high level.

Ordinarily, both halves of the output waveform are equal. That is, each transistor is held on for the same amount of time as the other. However, this is not always the case. In some applications it may be desirable to have the output at one state for a somewhat longer time in each cycle. The signal is then referred to as a rectangle wave. Several typical rectangle waves are illustrated in Fig. 6-13.

The 555 timer is often used as the basis for astable multivi-

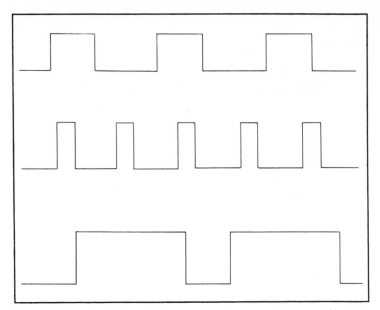

Fig. 6-13. Some astable multivibrators generate rectangle waves of various widths.

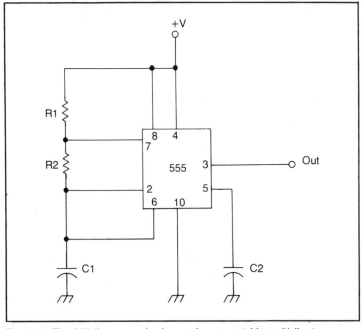

Fig. 6-14. The 555 timer can also be used as an astable multivibrator.

73

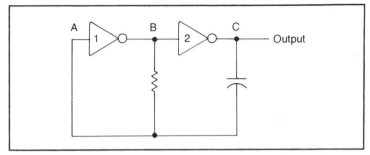

Fig. 6-15. A digital astable multivibrator can be constructed from two inverters.

brators in digital circuits, even though it is actually a linear device. Figure 6-14 shows the basic 555 astable multivibrator circuit. The frequency, or rate of cycle repetition, is determined by this formula:

$$F = \frac{1}{0.693 \ (R1 + 2R2) \ C1}$$

where F is the output frequency in hertz (cycles per second), while R1 and R2 are given in megohms and C1 is in microfarads.

One minor restriction to this circuit is that it cannot produce a true square wave with equal high and low output times. This is because the output high time is determined by both resistors and the capacitor:

$$T_H = 0.693 \ (R1 + R2) \ C1$$

while the low output time is controlled by just R2 and C1:

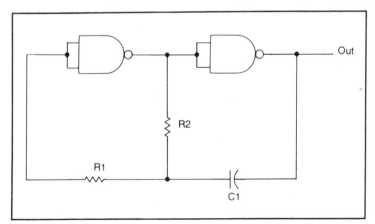

Fig. 6-16. A pair of NAND gates can be used as an astable multivibrator.

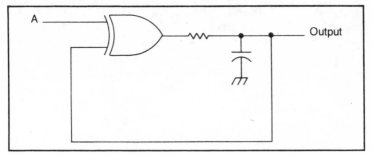

Fig. 6-17. An astable multivibrator can also be built around a single Exclusive-OR gate.

$$T_L = 0.693 \ (R2) \ C1$$

However, you can come very close to a square wave if the value of R1 is small with respect to the value of R2. Capacitor C2 will generally have a constant value of about 0.01 μF.

As with most 555 based circuits, this astable multivibrator can be compatible with most of the major logic families (see Chapter 12).

Astable multivibrators can also be constructed from digital gates, as illustrated in Figs. 6-15, 6-16, and 6-17. In Fig. 6-15, two inverters are combined with a single resistor and capacitor to create a simple astable multivibrator. To understand how this circuit works, let's assume we are starting out with a logic 0 at the point marked A in the diagram. Inverter 1 changes this to a logic 1 at point B, and inverter 2 changes the signal back to a logic 0 at point C, and this is fed to the output. While this is simple enough, the resistor and capacitor form a feedback network, which forces point A to logic 1 after a definite period determined by the RC time constant. The output will change back and forth between logic states at a rate that can be found via this formula:

$$F = \frac{1}{1.4 \ RC}$$

Because of component tolerances, this equation is something of an approximation.

This circuit is capable of a wide range of frequencies. The timing capacitor may take any value between about 0.01 μF and 10 μF.

A similar circuit is shown in Fig. 6-16. In this case NAND gates

75

are used in place of the inverters. Both inputs of each gate are tied together, so the gate functions essentially as an inverter. This circuit also includes a second resistor, but the basic operation of the circuit is essentially the same as the previous circuit.

The frequency formula for this astable multivibrator circuit is:

$$F = \frac{R2\ C1}{2.2}$$

Once again, this formula is basically an approximation. Resistor R1 should have a value that is about five to ten times the value of R2.

A different approach to a digital astable multivibrator is shown in Fig. 6-17. This circuit is built around a single Exclusive-OR gate. Input A should be held constant—tied permanently to either logic 0 or logic 1. Of course, the frequency is set by the resistor and capacitor.

In digital electronics circuits, astable multivibrators are commonly referred to as clocks, since their primary application is in the area of timekeeping, and synchronization (keeping the various portions of the circuit in step with each other).

Some applications require multiple-phase clocks. A multiple-phase clock generates two or more outputs that are out of phase with each other by a fixed amount. A very simple approach to creating a two-phase clock is shown in Fig. 6-18. Output 2 will always be exactly 180° out of phase with output 1. Many of the devices and circuits described in the following chapters will require clocks of some kind.

SCHMITT TRIGGERS

Closely related to the multivibrator is the Schmitt trigger. This is a circuit that can be used to clean up noisy signals into clean,

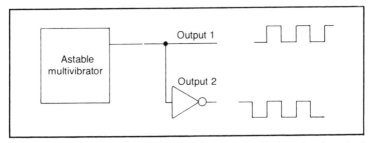

Fig. 6-18. A super simple two-phase clock can be created by inverting the output of an astable multivibrator.

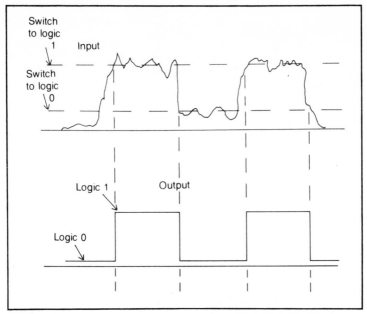

Fig. 6-19. A Schmitt trigger can be used to clean up noisy digital signals.

fast rising and falling digital pulses. The output of a Schmitt trigger goes high (logic 1) when the input signal is greater than a specific voltage, and it goes low (logic 0) when the input voltage drops below a given point.

The switch to the logic 0 point is usually lower than the switch to the logic 1 level, leaving a dead zone between the two. This makes the Schmitt trigger relatively insensitive to noise, as illustrated in Fig. 6-19. Dedicated Schmitt trigger ICs are available in most of the major logic families (see Chapter 12).

Chapter 7

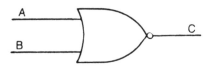

Flip-Flops

One of the most important types of subcircuits in digital electronics is the bistable multivibrator, or flip-flop. As we will soon see, this is a circuit with a memory. The bistable multivibrator was discussed briefly in the previous chapter. It is a multivibrator that can have a stable output at logic 0 or at logic 1, depending on how it was last triggered. There are a number of different types of flip-flops. Most can be constructed from NAND gates.

THE R-S FLIP-FLOP

The most basic type of flip-flop is the R-S flip-flop. This type of flip-flop has two inputs—set (S) and reset (R). Figure 7-1 shows how two NAND gates can be combined to act as a R-S flip-flop. Notice that there are two inputs (S and R) and two outputs (Q and $\overline{Q}$, or NOT Q). By definition, the Q and $\overline{Q}$ outputs should always have opposite values. That is, when Q is at logic 0, $\overline{Q}$ should be at logic 1, and when Q is at logic 1, $\overline{Q}$ should go to logic 0.

NAND gate *a* combines input R, and the output of NAND gate *b*, or output $\overline{Q}$. Conversely, NAND gate *b* works with input S, and the output of NAND gate *a*, or output Q.

Let's assume, both input S and input R start out at logic 0. No matter what a logic 0 is NANDed with, the output will always be a logic 1, because the NOT AND condition is true. The output of a NAND gate, as you should remember, is a logic 0 if, and only if, all of its inputs are at logic 1. If at least one input is at logic 0, the output

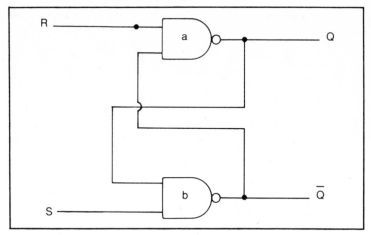

Fig. 7-1. A R-S type flip-flop can be made of two NAND gates.

will be a logic 1. Therefore, in this circuit, a logic 0 at both inputs will place a logic 1 at both outputs. Notice that both Q and $\overline{Q}$ are equal. This is a meaningless, and therefore undefined state for this circuit.

Now, let's feed a logic 1 to input S (set). Since both output Q and input S are now at logic 1, the output of NAND gate b (and therefore output $\overline{Q}$) drops to logic 0. The output of NAND gate a remains unaffected. In other words, when input S = 1, and input R = 0, then output Q = 1 and $\overline{Q}$ = 0. We say that the flip-flop is set.

Of course, reversing the inputs will give us the opposite results. When input S = 0 and input R = 1, output Q will go to logic 0, while output $\overline{Q}$ will go to logic 1. In this case, we say the circuit is reset.

If both inputs are at logic 1, the circuit will tend to get confused, and the output is unpredictable. Once again, we have an undefined state. The truth table for a R-S flip-flop is given in Table 7-1.

THE TOGGLE FLIP-FLOP

Suppose we take a R-S flip-flop, and add two more NAND gates connected as shown in Fig. 7-2. We now have a circuit known as a toggle flip-flop. It is driven by a single input. This input is generally labeled "clock," but it does not necessarily have to be fed by an astable multivibrator. Any logic signal can be used to trigger the circuit.

Let's assume output Q is at logic 1, and output $\overline{Q}$ is at logic 0,

Inputs		Outputs		
R S		Q $\overline{Q}$		
0 0		1 1	(disallowed state)	
0 1		1 0		
1 0		0 1		
1 1		? ?	(disallowed state)	

while a logic 0 is being fed to the input. Gate C NANDs the input with output $\overline{Q}$. In this case both inputs are 0's, so the output of gate C is a logic 1. This is the R input, which is fed into gate a, along with output $\overline{Q}$. NANDing 1 (R) and a 0($\overline{Q}$) again gives us a logic 1 output, holding output Q at its current value.

Gate d, on the other hand combines the input (0) with output Q (1) in a NAND operation. The output, of course, is a logic 1, since both inputs to the gate are not at logic 1. The output of gate d serves as the S input. S is NANDed with output Q by gate b. Both S and Q are now equal to 1, so the output of gate b (and therefore output $\overline{Q}$) must be a 0. A logic 0 fed to the input of this circuit causes absolutely no change in the output states. The circuit essentially "remembers" its previous state.

Now lets feed a 1 to the clock input. Gate d NANDs clock (1) and Q (1), to produce a 0 at S. This value is then NANDed by gate b with Q (1) to produce a logic 1 at output $\overline{Q}$.

Output $\overline{Q}$ is combined with the clock input by NAND gate c. Both of these signals are now at logic 1, so the output of this gate (R) is now a logic 0. Gate a performs a NAND operation on R and output

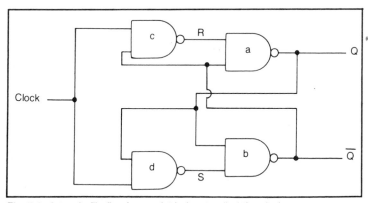

Fig. 7-2. A toggle flip-flop has a single input called the clock.

$\overline{Q}$. Since R = 0, and $\overline{Q}$ = 1, the result of this operation, and therefore output Q, is a logic 0. The 1 at the input caused the outputs to reverse states.

Now, if we let the input drop back down to logic 0, what will happen? Absolutely nothing. Output Q will remain at logic 0, and output $\overline{Q}$ will remain at logic 1. I will leave it to you to trace this out yourself.

Similarly, if the input changes again to a logic 1, R will be forced to 0 (R = Clock NAND $\overline{Q}$). Gate *a* NANDs R and $\overline{Q}$ for a logic 1 at output Q.

S equals Clock NAND Q, so in this case 1 (C) NAND 0 (Q) = 1. S is NANDed with Q by gate *b* to force $\overline{Q}$ to 1.

Each and everytime the clock input changes from logic 0 to logic 1, the outputs (Q and $\overline{Q}$) reverse states. The two outputs are always of opposite logic values. There are no disallowed states, since there are only two possible input conditions. If Clock = 0, the outputs remember their previous values. If Clock = 1, the outputs reverse values. This is summarized in Table 7-2. This circuit takes its name from the fact that the outputs toggle back and forth, each time the circuit is triggered by the clock input.

THE J-K FLIP-FLOP

If we use three-input NAND gates instead of the two-input gates shown in Fig. 7-2, we can create yet another type of flip-flop. This is the J-K flip-flop, and it is illustrated in Fig. 7-3.

Notice that there are five inputs to this circuit. They are Clock, S (set), R (reset), J, and K. The letters J and K do not stand for anything in particular, but they are in standard usage. The origins of this terminology is unclear.

If input S is at logic 1, and input R is at logic 0, output Q will be a logic 1, and output $\overline{Q}$ will be a logic 0. Inputs Clock, J, and K will be ignored. Input S is sometimes referred to as the "preset" input,

Table 7-2. This Truth Table Shows How a Toggle
Flip-Flop Reverses Output States with Each Input Pulse.

Clock Input	Previous Values of Outputs		New Values for Outputs	
	Q	$\overline{Q}$	Q	$\overline{Q}$
0	0	1	0	1
1	0	1	1	0
0	1	0	1	0
1	1	0	1	1

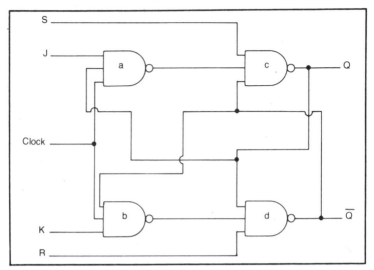

Fig. 7-3. The J-K flip-flop is an extension of the basic R-S flip-flop.

because it can force the circuit into the set (Q = 1) condition.

Similarly, if the R input is at logic 1, and the S input is at logic 0, the Q output will go to logic 0, regardless of the signals at the Clock, J, and K inputs. This input is variously called "reset," "clear," or "preclear," because it forces the circuit out of its set condition. Having both the S and R inputs at logic 0 is a disallowed state, and should be avoided. The output is not predictable under such conditions.

Inputting a logic 1 at both S and R enables the other inputs. The J and K inputs only take effect when a trigger signal is present at the Clock input. This allows the output to be held constant, even if the inputs to J and K are changed.

If both the J and K inputs are at logic 0 when a clock pulse is received, there will be no change in the output. Both outputs will retain their previous logic state.

If J equals logic 1 and K is at logic 0 when a clock pulse is received, output Q will become a logic 1, and output $\overline{Q}$ will become a logic 0, regardless of their previous values. The J input is sometimes labeled "set" (as opposed to "preset" or S).

If J is at logic 0 and K is at logic 1 when the Clock input is triggered, we have the opposite results. Output Q goes to logic 0 and output $\overline{Q}$ becomes a logic 1, regardless of their previous condition. The K input is occasionally labeled "reset" or "clear" (as opposed to "preclear" or R).

Table 7-3. The Truth Table for a J-K Type Flip-Flop Demonstrates Its Versatility.

Inputs					Output	
S	R	J	K	Clock (*)	Q	$\overline{Q}$
0	0	x	x	x	disallowed state — do not use	
0	1	x	x	x	0	1
1	0	x	x	x	1	0
1	1	x	x	0	no change	
1	1	0	0	1	no change	
1	1	0	1	1	0	1
1	1	1	0	1	1	0
1	1	1	1	1	outputs reverse previous value	

x = don't care
(*) for the Clock input

0 = no trigger pulse
1 = trigger pulse received

If both J and K are fed logic 1's, the circuit will behave like a toggle flip-flop. Each time a trigger pulse is received, outputs Q and $\overline{Q}$ will reverse values.

Thanks to its various operating modes, the J-K flip-flop is an extremely useful device, and is widely employed in digital circuits. It is probably the most widely used type of flip-flop. The various operating modes of the J-K flip-flop are outlined in Table 7-3.

THE D-TYPE FLIP-FLOP

If we combine a J-K type flip-flop with an inverter, as illustrated in Fig. 7-4, the J and K inputs will always be at opposite values. For example, if a logic 1 is fed into the input, this value will be passed directly to the J input, while the inverter converts it to a

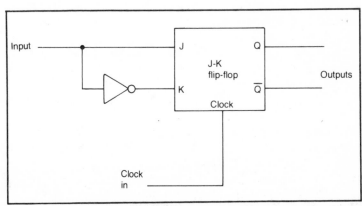

Fig. 7-4. A D-type flip-flop can easily be made from a J-K flip-flop.

83

logic 0 for the K input. As explained in the last section, when J = 1, and K = 0 and the Clock input is triggered, output Q becomes a logic 1, and output $\overline{Q}$ goes to logic 0, regardless of their previous states.

The other possible input is a logic 0. In this case, input J is at logic 0, and input K is the inversion, or logic 1. When the circuit is triggered, output Q is reset to 0, and output $\overline{Q}$ becomes a logic 1, as explained in the previous section.

In other words, when the input is at logic 1, and the circuit is triggered, output Q is at logic 1, and when the input is at logic 0 and a trigger pulse is received, output Q is at logic 0. Each time a clock pulse is fed into this circuit, the value at the input is assigned to the Q output. The $\overline{Q}$ output, of course, always takes on the opposite value.

What is the use of this? This circuit will remember the input as long as desired, until another pulse is fed to the Clock input (or power is removed from the circuit). This device is an ideal building block for storing data.

The input to this kind of flip-flop is usually referred to as D (for Data). The circuit is called a D-type flip-flop. The truth table, which is given in Table 7-4, demonstrates the simplicity of operation of this circuit.

Some D-type flip-flops also include S and R inputs, which behave exactly like the S and R inputs of a J-K type flip-flop. A complete circuit for a D-type flip-flop made up of NAND gates is shown in Fig. 7-5.

APPLICATIONS

Flip-flops have many applications in digital electronics. They are second in importance only to gates. One area of application for flip-flops has already been touched upon. Flip-flops can "re-

Table 7-4. Truth Table for a D-Type Flip-Flop.

Inputs		Outputs	
D	CLOCK	Q	$\overline{Q}$
x	0	no change	
0	1	0	1
1	1	1	0

x = don't care
for CLOCK input

0 = no trigger pulse
1 = trigger pulse received

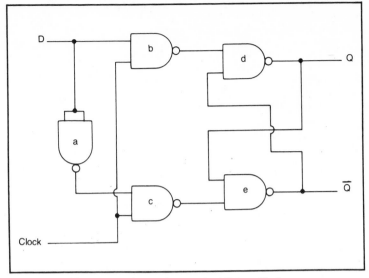

Fig. 7-5. NAND gates can be used to create a D-type flip-flop.

member" digital signals. Large scale memory circuits can be constructed from a collection of flip-flops. Digital memories will be discussed in detail in Chapter 11.

On a somewhat smaller scale, flip-flops can serve as signal latches, holding a signal long enough for it to be detected and used by other circuitry. This can be especially important in high speed digital circuits, where a signal may appear for only a tiny fraction of a second, and may be gone before a slower acting subcircuit has time to recognize and react to it.

Toggling a flip-flop can be used as a method of frequency division. To demonstrate this, let's start with a toggle flip-flop circuit, being driven by a regular clock. An astable multivibrator would be used to generate a string of clock pulses at a specific frequency.

When we start, let's say the Q output is at logic 0. The Clock input will also start at logic 0. A logic 0 clock signal has no effect on the output, so Q remains at logic 0. When the clock signal goes to logic 1, the Q output is toggled to its opposite state—logic 1. The clock then goes to logic 0. Again, there is no change in the output—it remains at logic 1. The next time the clock goes to logic 1, the output is toggled back to logic 0, and the pattern is repeated. If we write down the input and output signals for several cycles, we get something like this:

Clock Input	Q Output
0	0
1	1
0	1
1	0
0	0
1	1
0	1
1	0
0	0
1	1
0	1
1	0

Notice how it takes two input cycles for each output cycle. This is illustrated in Fig. 7-6. The output changes only half as fast as the input. In other words, the output runs at half the frequency as the input signal. The flip-flop effectively divides the input frequency by two.

Multiple flip-flops can be cascaded to divide by larger amounts.

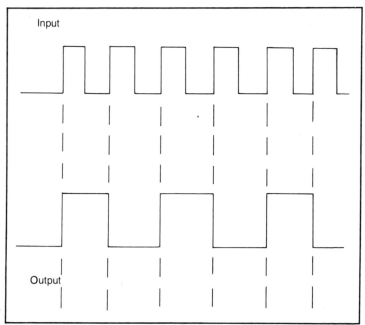

Fig. 7-6. A toggle flip-flop can divide the input frequency by two.

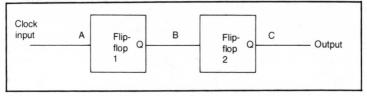

Fig. 7-7. Flip-flops can be cascaded to divide by four, or higher numbers.

For example, cascading two flip-flops, as shown in Fig. 7-7 will divide the input signal by four. This action is summarized in Table 7-5. Another way to look at this is that the two flip-flops count to four during each output cycle. Counter circuits will be examined in Chapter 8.

Flip-flops are also useful in performing mathematical operations such as multiplication and division of multiple-digit binary numbers. This is done by combining several flip-flops into a circuit called a shift register. There are also a number of other applications for shift registers. Several different types of shift registers have been developed for various applications. Shift registers will be discussed in detail in Chapter 9.

These are just a few of the most basic applications of flip-flops. Each of these applications, in turn, can be used in a multitude of ways in complex digital circuits.

CLOCKED LATCHES FOR SYNCHRONIZATION

In examining digital devices such as gates and flip-flops, we

Table 7-5. Two Cascaded Flip-Flops Can Divide the Input Frequency By Four (See Fig. 7-7).

Input		Output
A	B	C
0	0	0
1	1	1
0	1	1
1	0	1
0	0	1
1	1	0
0	1	0
1	0	0
0	0	0
1	1	1
0	1	1
1	0	1
0	0	1
1	1	0
0	1	0
1	0	0

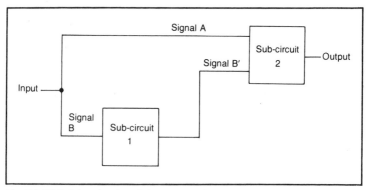

Fig. 7-8. Digital signals take a finite amount of time to pass through digital circuits, and this can cause synchronization problems in complex circuits.

tend to think of them as operating instantaneously. That is, as the input is received, the output goes to its appropriate state. Of course, in practical circuits the action is not quite instantaneous. It takes a finite amount of time for the circuitry to react. This time is extremely short by human terms, but it may be of significance in a digital circuit.

For example, let's assume a 1 MHz (1,000,000 Hz) signal is being passed through a complex digital circuit. At some point it is split off into two paths. Some of the signal is fed through a series of various gates before it is recombined with the other portion of the signal, which is passed directly to this point. This is shown in the block diagram of Fig. 7-8.

Signal B passes through subcircuit 1 in 0.000025 seconds, which is admittedly pretty fast. But the input signal is changing every 0.000001 seconds. Signal B' is 25 pulses behind signal A. The solution is to use a latch, or flip-flop, to store signal A for 25 cycles. The various circuits would be clocked to keep the signals in step with each other. We'll explore this idea in a bit more depth in the chapter on shift registers (Chapter 9).

Chapter 8

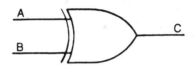

Counters

In many digital circuits, it is necessary to keep track of the number of pulses passing through some specific portion of the circuit. This task can be performed with special subcircuits called counters. There are many different types of counters in use today, but they all tend to work along the same lines.

COMBINING FLIP-FLOPS TO MAKE A BINARY COUNTER

A counter circuit is usually created by cascading a number of flip-flops. Of course, since flip-flops can be made from digital gates (as explained in the previous chapter), a counter circuit could be constructed directly from simple gates.

Figure 8-1 shows a very simple counter circuit. Four toggle flip-flops are connected in series, so the output of the first flip-flop triggers the second flip-flop, the output of the second triggers the third, and so forth.

Let's trace the action of this circuit. To start out, we will assume all of the flip-flops are cleared, or reset. That is, few outputs are at logic 0. When the first pulse is received, it triggers the first flip-flop from 0 to 1. The other flip-flops are left unchanged. The next pulse triggers flip-flop A back to logic 0, which in turn triggers the next flip-flop in line (B in this case) from 0 to 1. On the next pulse, only flip-flop A is affected as it is toggled from 0 to 1. On the fourth pulse, flip-flop A is triggered back to 0. The changing output of flip-flop A triggers flip-flop B, causing it to go from 1 back to 0.

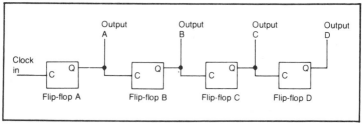

Fig. 8-1. A four-stage asynchronous counter can be constructed from flip-flops.

This, in turn, triggers flip-flop C from 0 to 1. This sort of thing continues for additional pulses, until all four outputs are at logic 1. This represents the maximum count of the device. On the next input pulse, all of the flip-flops are triggered back to 0, and the counter is once again cleared, and ready to begin a new counting cycle.

The counting sequence for a four flip-flop counter such as the one shown in Fig. 8-1 is outlined in Table 8-1. Notice how the outputs count in binary from 0000 to 1111 (decimal 15), then the cycle is repeated. We can call this circuit a 16-step counter. Another

Table 8-1. This Table Demonstrates the Counting Cycle of a Four-Stage Binary Counter.

Input Pulse #	Outputs D	C	B	A	Count Value (in decimal)
0	0	0	0	0	0
1	0	0	0	1	1
2	0	0	1	0	2
3	0	0	1	1	3
4	0	1	0	0	4
5	0	1	0	1	5
6	0	1	1	0	6
7	0	1	1	1	7
8	1	0	0	0	8
9	1	0	0	1	9
10	1	0	1	0	10
11	1	0	1	1	11
12	1	1	0	0	12
13	1	1	0	1	13
14	1	1	1	0	14
15	1	1	1	1	15
16	0	0	0	0	0
17	0	0	0	1	1
18	0	0	1	0	2
19	0	0	1	1	3
20	0	1	0	0	4

and so forth . . .

common way to say the same thing is to call it a modulo-sixteen counter. The concept of modulo will be discussed later in this chapter.

If we add a fifth flip-flop stage, as shown in Fig. 8-2, we increase the potential count by a factor of two. In other words, we can now count up to the five digit binary number, 11111, which is equivalent to decimal 31. This would give us a 32-step counter.

While this simple approach is adequate for some applications, there are some problems. The primary problem is that each stage's output triggers the next stage. This means the outputs change sequentially rather than simultanously. Why should this matter? Well, let's assume the count is 0111, and a new pulse is received. Flip-flop A is triggered first, so for the brief instant it takes flip-flop B to respond to its new input, the outputs will be 0110. Similarly, once flip-flop B's output changes, it will take a very short but finite amount of time for flip-flop C to react, so the output will temporarily be 0100. The change in flip-flop C will also occur before flip-flop D has time to change. In other words, instead of the correct counting step:

0111

1000

we get a string of meaningless counts between the correct values:

0111	
0110	(flip-flop A changes)
0100	(flip-flop B changes)
0000	(flip-flop C changes)
1000	(flip-flop D changes)

Obviously this problem becomes more severe as the number of flip-flop stages is increased.

Whether or not intermediate incorrect counts will be a problem

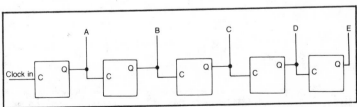

Fig. 8-2. A flip-flop counter can include as many stages as needed.

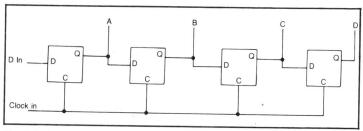

Fig. 8-3. A synchronous counter requires a separate clocking input.

depends on the application at hand. If the counter is just driving LED readouts, the incorrect counts will be far too brief to be visible, so they can simply be ignored. However, if the counter outputs are fed to additional digital circuitry, they will probably be accepted as valid counts, and that can obviously cause major problems in the operation of the system. After all, its rare that we'd actually want a circuit that counts 7 - 6 - 4 - 0 - 8.

Because the flip-flops are operated sequentially, rather than in synchronization, this type of counter is called an asynchronous counter. We can avoid the problems of the asynchronous counter if we replace the simple toggle flip-flops with D-type flip-flops, as

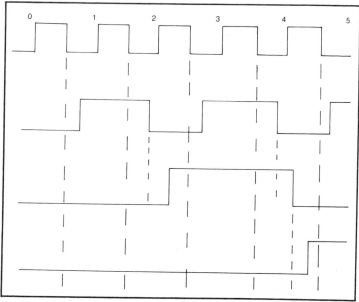

Fig. 8-4. These timing graphs show that an asynchronous counter outputs several incorrect counts before settling to its correct value.

92

illustrated in Fig. 8-3. Notice that there are now two inputs: the data (which is the string of pulses to be counted) and the clock, (which causes the flip-flops to check their inputs and change their output states as appropriate simultaneously).

To more clearly show the differences, Fig. 8-4 shows the input and output timing graphs for an asynchronous counter, while the comparable graphs for a synchronous counter (like Fig. 8-3) are shown in Fig. 8-5.

THE MODULO OF A COUNTER

The maximum count a counter circuit is capable of is referred to as the modulo of that counter. Counters with modulos equal to any whole number greater than 1 can be constructed using flip-flops, as described above.

For modulos that are powers of two ($2^2 = 4$, $2^3 = 8$, $2^4 = 16$, $2^5 = 32$, and so on), the design is quite simple. All that such a counter needs is a string of the appropriate number of flip-flops—one for each digit in the outputted binary number (or, for each power of two).

But what if we want a counter with a modulo of five? Five is not a power of two. Most flip-flops have a clear (or R or reset) input that

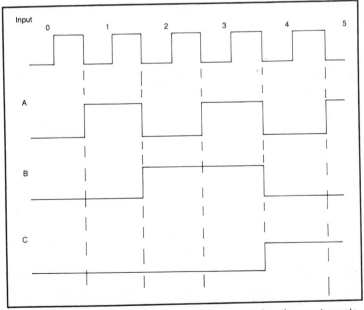

Fig. 8-5. A synchronous counter does not produce spurious incorrect counts.

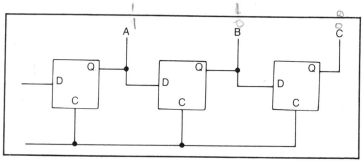

Fig. 8-6. Three flip-flops can be combined to make a modulo-eight counter.

can force the Q output back to logic 0. By using some digital gates, we can force the flip-flops to clear when a specific count is reached.

If we need a counter with a modulo of five, we want the output count sequence to run like this:

000
001
010
011
100
000
001

and so on. The output count needs to be forced back to 000 after a count of 100. To design such a counter, we first set up a modulo-8 counter, since that is the next highest power of two. This requires three flip-flops as shown in Fig. 8-6.

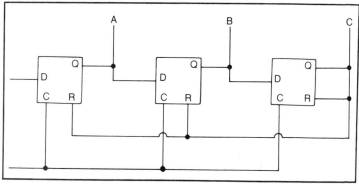

Fig. 8-7. A modulo-eight counter can easily be converted to a modulo-five counter by feeding back the four count.

We want the counter to reset back to zero after the output count goes to A = 0, B = 0, C = 1. This is the only valid output where output C is at logic 1, so we can feed this out back to the clear inputs of the flip-flops as shown in Fig. 8-7.

Even though the flip-flops are theoretically capable of counting to 111 (decimal 7), the feedback prevents the count from ever exceeding 100 (decimal 4). Eureka, we have a modulo-five counter!

Now let's change this to a modulo-six counter. This is only slightly more complex than the previous problem. In this case we want the output counts to cycle through this sequence:

$$000$$
$$001$$
$$010$$
$$011$$
$$100$$
$$101$$
$$000$$
$$001$$
$$010$$

and so on, resetting to 000 after every count of 101.

Now, we can not just use the C output as the feedback signal to clear the flip-flops, or the circuit will never get a chance to output a count of 101. We need to reset the flip-flops when both output A and output C are at logic 1. (Output B doesn't really matter, since the

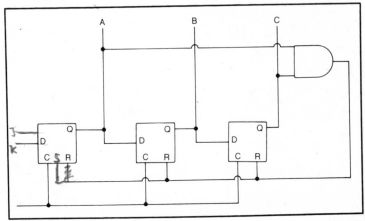

Fig. 8-8. An AND gate is used to convert a modulo-eight counter to a modulo-six counter.

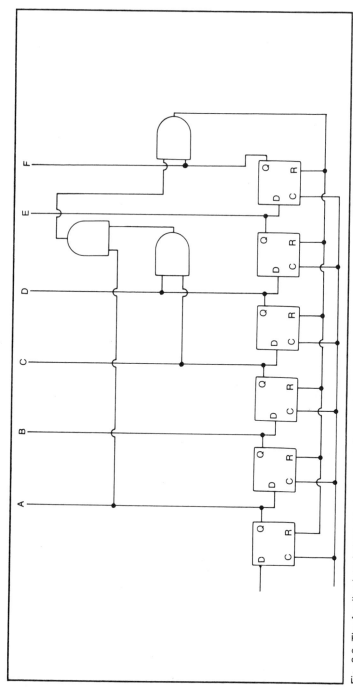

Fig. 8-9. The feedback technique can be expanded as much as neccessary. Here we have a modulo-46 counter.

count of 111 will never be reached.) Clearly, we should call in an AND gate. Outputs A and C are ANDed and the result is used to clear the counter. This circuit is shown in Fig. 8-8.

By using the right combination of gates, literally any whole number modulo can be achieved. Some are slightly more complex than the examples we have described here. For instance, consider the gating required for the modulo-46 counter shown in Fig. 8-9. The counter is cleared after the output count passes 101101.

In most cases AND gates can be used to set the modulo feedback value. The gates should select a count that is one less than the modulo desired.

The gates required for modulos of 3 through 15 are illustrated in Fig. 8-10. Modulo four and modulo eight are omitted here, since they are powers of two, and can be generated directly from two or three flip-flops respectively, without forced resetting.

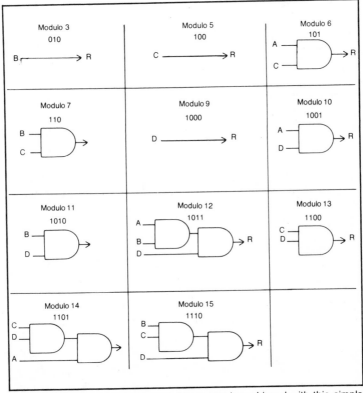

Fig. 8-10. Modulos of three through fifteen can be achieved with this simple gating arrangement.

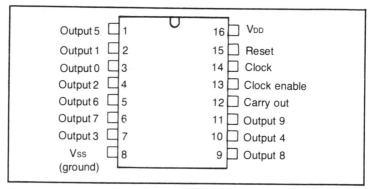

Fig. 8-11. Many counter circuits are available in IC form, such as the CD4017.

Remember, the gates are arranged to reset the flip-flops after the count reaches a value one less than the desired modulo. This is because 0000 is considered a count. Many counters are available in IC form. Generally, they can be used in the same way as the flip-flop counters already described.

Some IC counters are designed to count out in a somewhat different way. As an example, we will consider the CD4017

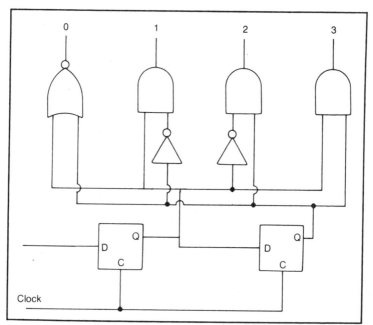

Fig. 8-12. Flip-flops and gates can be combined to create a one-of-four counter.

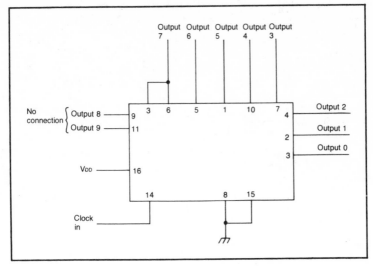

Fig. 8-13. This CD4017 circuit will count to 7 and then stop.

counter/divider chip. This is a CMOS device (see Chapter 12). The pinout diagram for the CD4017 is given in Fig. 8-11.

This IC has ten outputs, numbered from 0 to 9. On any specific count, only one of the outputs is high (logic 1), and the other nine outputs are low (logic 0). We can simulate this with flip-flops and gates as shown in the four-step counter illustrated in Fig. 8-12. Only one output can be at logic 1 at any time.

By grounding pin 15 of a CD4017 and connecting pin 13 to one of the outputs, the counter will count from 0 to that output's value, and then stop. Figure 8-13 shows how this device can be wired to count to 7 and then stop. The circuit may be reset by temporarily disconnecting pin 15 from ground and touching it to a positive voltage source (generally the V_{DD} supply will be used).

A slightly different application is illustrated in Fig. 8-14. Here the connections to pins 13 and 15 are reversed from Fig. 8-13. This time pin 13 is grounded and pin 15 is connected to the maximum count output. If the circuit is wired as shown here, the outputs will count from 0 to 7, then revert back to 0 and start over. In other words, the outputs will go to logic 1 in this order— 0, 1, 2, 3, 4, 5, 6, 7, 0, 1, 2, 3, 4, 5, 6, 7, 0, 1, 2, 3, and so on. Of course, any count up to nine may be setup in this manner.

But what if you need a circuit that will count higher than nine? Counter circuits like the CD4017 can be cascaded to create higher counts. For example, cascading two CD4017's as shown in Fig. 8-15

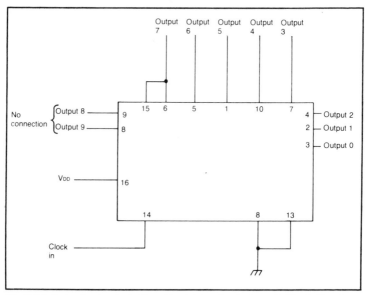

Fig. 8-14. A simple change in the wiring of the circuit shown in Fig. 8-13, allows the count to repeatedly recycle.

will allow you to set up counts up to 99. Counting 0, this gives you a modulo of 100.

One counter is used to represent the tens column, while the other is the ones column. Since two outputs are combined to make up each number, a two-input AND gate will be needed to setup the

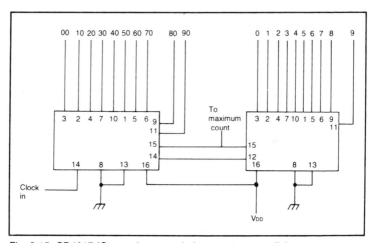

Fig. 8-15. CD4017 ICs can be cascaded to create a two-digit counter.

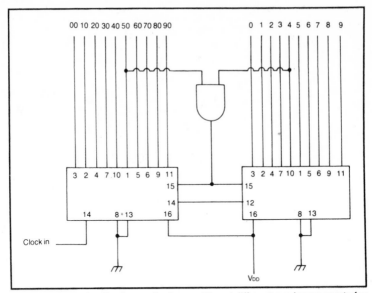

Fig. 8-16. Gated feedback can also be used to set different maximum counts for the CD4017. The circuit shown will count to 54 then recycle.

desired maximum count. The circuit shown in Fig. 8-16 will count up to 54, then recycle and repeat.

This system can be easily expanded for still higher counts. A third CD4017 will allow counts up to 999. A fourth chip will extend the maximum possible count to 9999.

BCD

While the binary numbering system, or one of its derivitives (octal or hexadecimal), is highly convenient for digital electronics circuits, it is rather awkward for human beings to deal with. Often it is desirable to provide for automatic conversion between the binary and the more familiar decimal systems. The circuitry required to perform this conversion (in either direction) is referred to as *binary coded decimal* circuitry, or BCD for short.

In the BCD system, four binary digits are grouped to represent each decimal digit. This is similar to the hexadecimal system described in Chapter 1, but only up to a point. A four digit binary number can consist of any of sixteen combinations. The first ten are used to represent the decimal digits 0 through 9:

$$0000 = 0$$
$$0001 = 1$$

$$0010 = 2$$
$$0011 = 3$$
$$0100 = 4$$
$$0101 = 5$$
$$0110 = 6$$
$$0111 = 7$$
$$1000 = 8$$
$$1001 = 9$$

This much is perfectly straightforward. But what of the remaining six combinations?

$$1010 = ?$$
$$1011 = ?$$
$$1100 = ?$$
$$1101 = ?$$
$$1110 = ?$$
$$1111 = ?$$

None of these combinations have a single digit equivalent in the decimal system. Therefore, in BCD, these six combinations are disallowed states, and are omitted.

To express a two-digit decimal number, such as 13, two groups of four binary digits are required. The BCD equivalent for decimal 13 is 0001 0011. Each decimal digit is expressed separately. Contrast this with the ordinary binary system. Decimal 13 equals binary 1101.

BCD values greater than nine require more binary digits than straight binary numbers. This can be especially a problem when relatively large numbers like decimal 3,754 must be stored. In straight binary, this value works out to 111010101010 (12 digits), while in BCD it becomes 0011 0111 0101 0100 (16 digits).

However, the BCD format is much easier to convert directly into its decimal equivalent. Finding the decimal equivalent for binary 1001100010 would require some tedious mathematics. On the other hand, if the same value is expressed in BCD, we can easily solve the problem a digit at a time. The BCD value is 0110 0001 0000, so the first digit (0110) has a value of 6, the second digit (0001) is 1, and the last digit (0000) is 0. The decimal value is 610.

BCD circuitry is especially useful when numerical displays like seven-segment LEDs (or LCDs) are used for the output. A seven-segment display consists of seven LED (or LCD) elements arranged in a figure-eight pattern. Any single digit can be displayed on

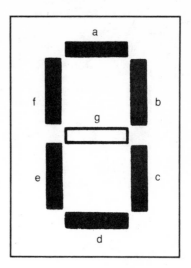

Fig. 8-17. Lighting segments a, b, c, d, e, and f on a seven-segment display will produce a 0.

such a unit, as illustrated in Figs. 8-17 through 8-26. Multidigit numbers can be displayed by using more than one seven-segment unit.

Seven-segment displays are sometimes used for hexadecimal values. The additional digits for the hexadecimal system (A, B, C, D, E, F) can be displayed as shown in Figs. 8-27 through 8-32. However, the human operator must then mentally convert the value to the decimal system, so using BCD circuits to display the output directly in decimal form is preferable.

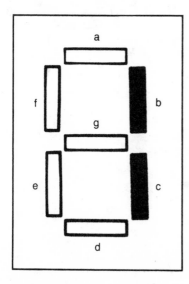

Fig. 8-18. A 1 can be displayed on a seven-segment display by lighting segments b and c.

103

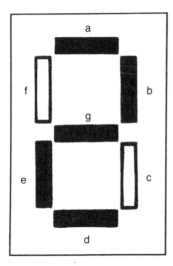

Fig. 8-19. Segments a, b, d, e, and g must be lit to display a 2.

Fig. 8-20. Lighting segments a, b, c, d, and g displays a 3.

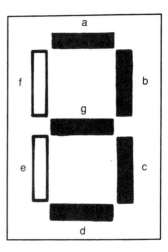

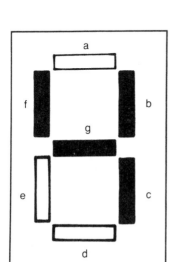

Fig. 8-21. Segments b, c, f, and g are lit to display a 4.

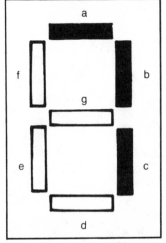

Fig. 8-22. A 5 can be displayed by lighting segments a, c, d, f, and g.

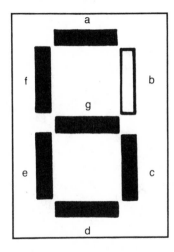

Fig. 8-23. A 6 can be displayed by lighting segments a, c, d, e, f, and g.

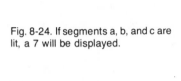

Fig. 8-24. If segments a, b, and c are lit, a 7 will be displayed.

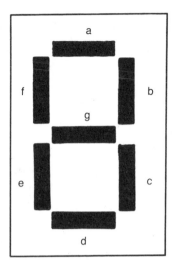

Fig. 8-25. Segments a, b, c, d, e, f, and g are all lit to display an 8.

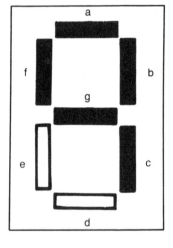

Fig. 8-26. To display a 9 segments a, b, c, f, and g are lit.

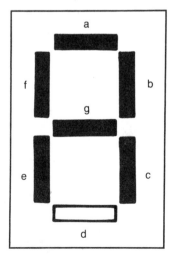

Fig. 8-27. A hexadecimal A can be displayed by lighting segments a, b, c, e, f, and g.

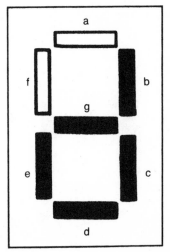

Fig. 8-28. For the hexadecimal digit B, a small b is created by lighting segments c, d, e, f, and g.

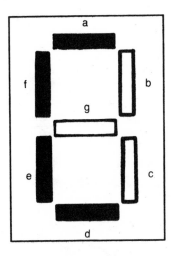

Fig. 8-29. A hexadecimal C is displayed by lighting segments a, d, e, and f.

Fig. 8-30. To avoid confusion with a 0, a hexadecimal D is displayed as a small d, by lighting segments b, c, d, e, and g.

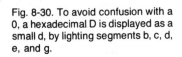

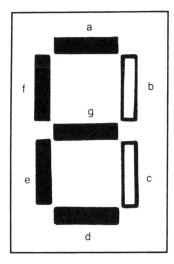

Fig. 8-31. Lighting segments a. d, e, f, and g displays a hexadecimal E.

The BCD system can also be used to input values. Special switching circuits convert the decimal digits fed into the device into the appropriate four digit binary equivalent.

Many BCD devices are available in integrated circuit form. The MC14553B is a three-decade BCD counter. This is a CMOS device (see Chapter 12). The MC14553B includes three counter stages, each with their own set of latches, so the count can be held, even in the absence of the input signal. Pins 9, 7, 6, and 5 are the four binary digit outputs. They are in BCD format, so the output will never exceed 1001 (decimal 9).

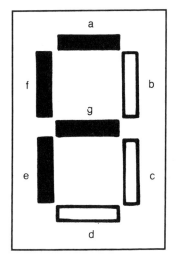

Fig. 8-32. A hexadecimal F can be displayed by lighting segments a, e, f, and g.

Table 8-2. The Output Sequence of a Three Decade BCD Counter Is Summarized in This Table.

BCD Output	Decimal Equivalent
0000 0000 0000	0
0000 0000 0001	1
0000 0000 0010	2
0000 0000 0011	3
0000 0000 0100	4
0000 0000 0101	5
0000 0000 0110	6
0000 0000 0111	7
0000 0000 1000	8
0000 0000 1001	9
0000 0001 0000	10
0000 0001 0001	11
0000 0001 0010	12
0000 0001 0011	13
0000 0001 0100	14
0000 0001 0101	15
0000 0001 0110	16
0000 0001 0111	17
0000 0001 1000	18
0000 0001 1001	19
0000 0010 0000	20
0000 0010 0001	21
0000 0010 0010	22
0000 0010 0011	23
0000 0010 0100	24
***	
0000 1001 0111	97
0000 1001 1000	98
0000 1001 1001	99
0001 0000 0000	100
0001 0000 0001	101
0001 0000 0010	102
0001 0000 0011	103
0001 0000 0100	104
0001 0000 0101	105
0001 0000 0110	106
0001 0000 0111	107
0001 0000 1000	108
0001 0000 1001	109
0001 0001 0000	110
0001 0001 0001	111
0001 0001 0010	112
0001 0001 0011	113
***	
1001 1001 0111	997
1001 1001 1000	998
1001 1001 1001	999
0000 0000 0000	0
0000 0000 0001	1
0000 0000 0010	2
0000 0000 0011	3
and so on . . .	(*** = skipped counts)

The three output values (or decades—ones, tens, and hundreds) all use the same output lines. Which one is currently being fed to the outputs is determined by pins 1, 2, and 15. In practical circuits, the three decade outputs will be rapidly switched between each other. An external latch system will direct the output to the appropriate display, or other circuit. Table 8-2 demonstrates how a BCD counter counts.

Chapter 9

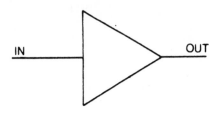

Shift Registers

Closely related to digital counters are a group of circuits known as shift registers. A binary number entered into a shift register, can be shifted place-by-place in either direction. This process is made clearer by direct example.

Let's assume that in each of the following examples we are starting out with the binary number 011001. Some shift registers will shift the digits one place to the left on each clock pulse. The newly open spaces on the right are filled with 0's. Digits shifted out of the left-most column are lost. In this case, the shift sequence will look something like this:

Clock pulse 0	011001	(original number)
Clock pulse 1	110010	
Clock pulse 2	100100	
Clock pulse 3	001000	
Clock pulse 4	010000	
Clock pulse 5	100000	
Clock pulse 6	000000	(The register is
Clock pulse 7	000000	now cleared, so
Clock pulse 8	000000	no further
Clock pulse 9	000000	changes will take
		place.)

Other shift register circuits will loop around the left-most digit to the right-most column as the number is shifted to the left, causing the digits to cycle through each position like this:

Clock pulse 0	011001	(original number)
Clock pulse 1	110010	
Clock pulse 2	100101	
Clock pulse 3	001011	
Clock pulse 4	010110	
Clock pulse 5	101100	
Clock pulse 6	011001	(the original
Clock pulse 7	110010	number again)
Clock pulse 8	100101	
Clock pulse 9	001011	

This pattern will be repeated indefinitely.

There are also shift register circuits that will shift the numbers to the right, instead of to the left. For example, a shift to the right and clear (incoming digits are set to 0, and outgoing digits are lost) would exhibit the following pattern:

Clock pulse 0	011001	(original number)
Clock pulse 1	001100	
Clock pulse 2	000110	
Clock pulse 3	000011	
Clock pulse 4	000001	
Clock pulse 5	000000	
Clock pulse 6	000000	(The register is
Clock pulse 7	000000	now cleared.)
Clock pulse 8	000000	
Clock pulse 9	000000	

Finally, some shift registers shift the numbers to the right and recycle. That is, as a digit is shifted out of the right-most column, it is replaced in the left-most position, like this:

Clock pulse 0	011001	(original number)
Clock pulse 1	101100	
Clock pulse 2	010110	
Clock pulse 3	001011	
Clock pulse 4	100101	
Clock pulse 5	110010	(the original
Clock pulse 6	011001	number again)

Clock pulse 7	101100
Clock pulse 8	010110
Clock pulse 9	001011

Once again, this pattern will be repeated indefinitely.

APPLICATIONS

At first glance, the action of a shift register may seem rather pointless, but it is actually an extremely useful device in digital electronics. Applications for shift registers include short term memories, mathematical operations, and digital delays.

When you multiply two multidigit numbers, you mentally perform a shifting function. For example, if you need to multiply 417 by 54, you break it down something like this:

$$(5 \times 7) + (5 \times 10) + (5 \times 400) \times 10$$
$$+ \ (4 \times 7) + (4 \times 10) + (4 \times 400) \ =$$
$$(35 + 50 + 2000) \times 10 + (28 + 40 + 1600)$$
$$(2085) \times 10 + 1668 = 20850 + 1668 =$$
$$22518$$

You probably don't perform all these steps consciously. Most people will write the problem down like this:

$$
\begin{array}{r}
417 \\
\times \ 54 \\
\hline
2085 \\
+ \quad 1668 \\
\hline
22518
\end{array}
$$

Notice that all of the steps must be performed at some level to solve the problem.

A digital electronics circuit goes through the same steps. A shift register allows each number to be examined digit by digit. Of course, shift registers come in handy for other mathematical operations besides multiplication.

Another important application for shift registers is short-term storage of binary numbers. In some circuits, different parts of the circuit may operate at different speeds. For example, let's say subcircuit A generates a digital signal at high speed. This signal must be fed to subcircuit B, which operates at a slower rate. The binary numbers can be stored in a shift register long enough for subcircuit B to recognize and properly manipulate each number.

SERIAL AND PARALLEL SIGNALS

Some circuits operate on one digit at a time. For instance, the binary number 100110 would be broken down into six separate signals, one after another:

$$
\begin{array}{c}
1 \\
0 \\
0 \\
1 \\
1 \\
0
\end{array}
$$

This type of operation is called serial operation, because the number is generated or received as a series of separate digits.

Serial operation is often desirable where digital signals must be transmitted from one circuit to a separate circuit. Only two wires are needed between the two circuits—one to carry the separate binary digits, and the other to serve as a common bus, or ground line. This is illustrated in Fig. 9-1.

On the other hand, many digital circuits operate in parallel fashion. In this case, binary numbers are transmitted as a unit, all of the digits appearing simultaneously. Obviously, this allows for faster operation, since a binary number like 100110 can be transmitted from one circuit to another in one step, rather than in the six separate steps required for parallel operation.

When connecting digital circuits for parallel operation, more interconnecting wires are needed. Each digit in the number must have its own transmission line. Since a common bus or ground line is also required, a parallel transmission system for numbers of x digits would call for x + 1 interconnecting wires. For example, a six

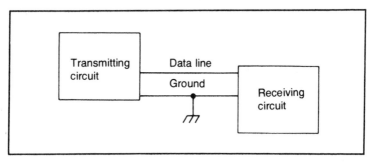

Fig. 9-1. Serial data transmission requires only two connecting wires between the transmitter and the receiver.

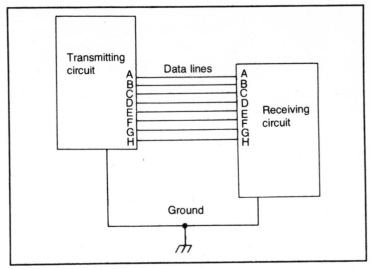

Fig. 9-2. Parallel data transmission demands a separate connecting wire for each bit, plus a common ground line.

digit system would need seven transmission lines, as illustrated in Fig. 9-2.

In many cases a shift register can convert serial numbers to the parallel format or vice versa. There are four basic types of shift registers, distinguished by the formats of their inputs and outputs. We will discuss each possible combination in the next few pages.

SERIAL-IN/SERIAL-OUT SHIFT REGISTERS

The first type of shift register we will examine, is also the simplest. Digital data is fed into the circuit in serial fashion (one binary digit, or bit, at a time), and after passing through each stage of the shift register, is fed to the output in serial form. This type of shift register is called a serial-in/serial-out shift register, for obvi-

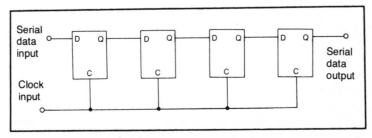

Fig. 9-3. A SISO (serial-in/serial-out) shift register can be made from flip-flops.

ous reasons. Sometimes this is abbreviated to SISO. A typical serial-in/serial-out shift register circuit is shown in Fig. 9-3.

Since the output of this circuit is identical to the input, such a circuit might not seem to be of much value. However, since each digit is moved forward through the circuit one stage for each controlling clock pulse, the output is delayed for a number of clock pulses equal to the number of stages. This is useful when the same digital number must be fed to two or more circuits with unequal propagation times (due to differing number of gates or flip-flop stages, or different logic families, or whatever). The faster subcircuit can have its input delayed by a given number of clock pulses with a SISO shift register. In other words, a serial-in/serial-out shift register can be employed as a digital delay, or temporary memory storage device.

SERIAL-IN/PARALLEL-OUT SHIFT REGISTERS

The next type of shift register accepts its input signals in bit-by-bit serial form, but feeds out the data in parallel format. This kind of circuit is called a serial-in/parallel-out shift register, or SIPO, for short.

As Fig. 9-4 clearly shows, a SIPO circuit is quite similar to a SISO circuit, except an output is taken off at each stage. To better understand how a serial-in/parallel-out shift register works, we will compare the outputs for a number of cycles in a typical system. We will assume the shift register in our example has four stages, and therefore, four parallel outputs. All stages are cleared (at logic 0) when we begin:

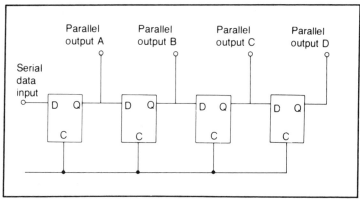

Fig. 9-4. A SIPO (serial-in/parallel-out) shift register can be made from a SISO shift register by adding outputs after each stage.

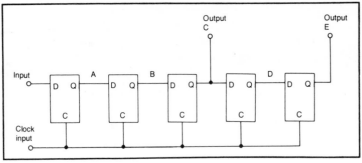

Fig. 9-5. A SIPO shift register can be used as a multiple digital delay.

Clock Pulse #	Serial Input	Outputs			
		A	B	C	D
0	0	0	0	0	0
1	1	0	0	0	0
2	1	1	0	0	0
3	0	1	1	0	0
4	1	0	1	1	0
5	0	1	0	1	1
6	1	0	1	0	1
7	0	1	0	1	0
8	0	0	1	0	1
9	0	0	0	1	0
10	0	0	0	0	1
11	0	0	0	0	0
12	0	0	0	0	0

One obvious application of the serial-in/parallel-out shift register is the conversion of serial data into parallel form. This type of shift register can also be used when digital delays of different lengths are required. Outputs are taken off of the appropriate stages. For example, let's say we need subcircuit B to be delayed 3 clock pulses (with respect to subcircuit A) and subcircuit C must be delayed 5 clock pulses (also with respect to subcircuit A). We can accomplish this with a three-stage SISO shift register and a separate five-stage SISO shift register. However, we can simplify the circuitry and lower the number of parts required (and therefore the system cost) by using a single five-stage SIPO shift register instead. Output C is used to drive subcircuit B and output E is fed to the input of subcircuit C. The other three parallel outputs (A, B, and D) are simply left unconnected. This is illustrated in Fig. 9-5.

PARALLEL-IN/SERIAL-OUT SHIFT REGISTERS

So far we have only looked at shift registers that accept their input data in serial form. Shift register circuits that accept parallel input data also exist. These circuits accept entire binary numbers in a single clock pulse.

Figure 9-6 shows a typical circuit for a parallel-in/serial-out shift register. The name is self-explanatory, of course. The input to this device is in parallel format, while the output is generated in serial form.

For example, if the binary number 100110 is fed into a parallel-input/serial-output shift register, the output for the next six clock pulses would look like this:

Clock Pulse #	Output
1	0
2	1
3	1
4	0
5	0
6	1

The least significant digit (the ones column) is fed to the output first.

A few parallel-in/serial-output shift registers work in just the opposite fashion. The most significant digit (highest value column) is outputted first. For the same parallel input (100110), the serial output would appear in this order:

Clock Pulse #	Output
1	1
2	0
3	0
4	1
5	1
6	0

Parallel-input/serial-output is generally abbreviated as PISO.

PARALLEL-INPUT/PARALLEL-OUTPUT SHIFT REGISTERS

There is one remaining input/output combination. This is

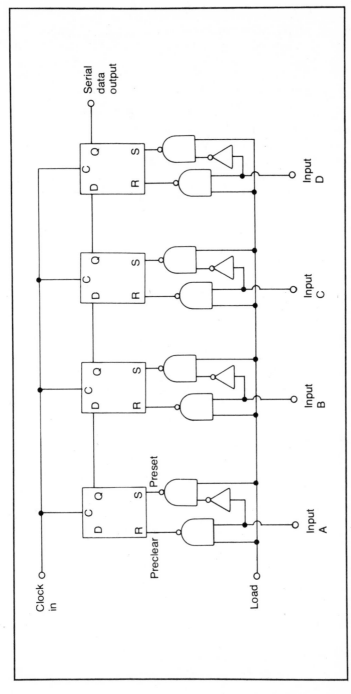

Fig. 9-6. A PISO (parallel-in/serial-out) shift register is more complex than either the SISO or SIPO type.

119

when both the input and the output are in parallel format. Not surprisingly, the circuit for this is called a parallel-input/parallel-output shift register. This can be shortened to PIPO. A parallel-input/parallel-output shift register circuit is illustrated in Fig. 9-7.

RING COUNTERS

Closely related to the shift register is the ring counter. A ring counter circuit includes a feedback path for the output of the last stage to be returned to the first stage. This causes a sequence of binary numbers that repeat in a regular sequence.

As an example, let's assume we have a ring counter that is loaded with the binary number 100110. If we follow the output for a number of clock pulses in a serial-out ring counter, we'd see this pattern:

Clock Pulse #	Output	
1	0	
2	1	
3	1	
4	0	
5	0	
6	1	
7	0	(The sequence begins to repeat here.)
8	1	
9	1	
10	0	
11	0	
12	1	
13	0	(The sequence starts to repeat again)
14	1	
15	1	

and so on.

Some ring counter circuits have parallel outputs. Starting with the same initial value (100110), the output pattern would look like this:

Clock Pulse #	Outputs
1	100110
2	010011

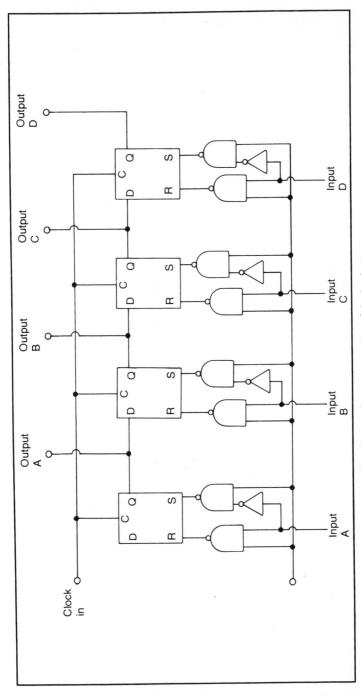

Fig. 9-7. A PIPO (parallel-in/parallel-out) shift register is the most complex of the basic types.

121

Clock Pulse #	Outputs	
3	101001	
4	110100	
5	011010	
6	001101	
7	100110	(the original number)
	010011	
9	101001	
10	110100	
11	011010	
12	001101	
13	100110	(the original number again)
14	010011	
15	101001	

and so on.

Since on each clock pulse each digit is moved over one place to the right (except the right-most digit which is looped around to the now vacant left-most positon), this is often called a shift-right operation.

Other circuits perform a shift-left operation, like this:

Clock Pulse #	Outputs	
1	100110	
2	001101	
3	011010	
4	110100	
5	101001	
6	010011	
7	100110	(the original number)
8	001101	
9	011010	
10	110100	

and so forth.

Notice that all of the same numbers appear whichever direction shift is used. The difference is in the order of their appearance.

Figure 9-8 shows the schematic diagram for a ring counter with both serial and parallel outputs. Like the other shift registers

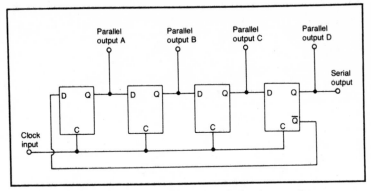

Fig. 9-8. A useful variation on the shift register is the ring counter.

discussed in this chapter, a ring counter may be built using flip-flops.

SHIFT REGISTER ICS

Many dedicated integrated circuits for performing shift register (and ring counter) functions are commercially available. Because such devices are somewhat more complex than gates or flip-flops, such ICs are generally considered to be MSI, or Medium-Scale Integration.

Chapter 10

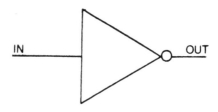

Other Digital Devices

A number of additional basic digital circuits are also in common use. This chapter will explore a few of the more important of these. Like most other digital circuits, these devices may be built from individual digital gates, but they are also available in prefabricated integrated circuit form. The ICs for performing the digital functions described in this chapter are MSI (Medium-Scale Integration) units.

MULTIPLEXERS

In some complex digital systems we will need one signal at a given point in the circuit part of the time, but other signals will be needed instead at other times. To build such a system, we obviously need some way to select between two or more possible inputs. A subcircuit that has been developed for this purpose is called a multiplexer. Sometimes the name is abbreviated as MUX.

Figure 10-1 shows how a simple multiplexer can be made from NAND gates. This circuit has four main inputs (labeled 1 through 4 in the diagram) and two control inputs (labeled A and B in the diagram). The logic signals fed to the control inputs determine which of the main input signals will reach the output. Only one of the main input lines is active at any one time. The truth table for this circuit is given in Table 10-1. Notice how only one of the four main input lines is of significance for any specific combination of values for control inputs A and B.

Because the control inputs can be used to select any of the main

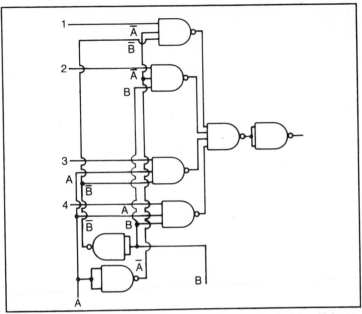

Fig. 10-1. Standard NAND gates can be combined to create a 1-of-4 multiplexer.

inputs for the output, this type of circuit is occasionally called a data selector, although multiplexer is the preferred name.

The circuit shown in Fig. 10-1 is a 1-of-4 multiplexer, since any one of the four main inputs may be selected. The same principle is commonly expanded to make 1-of-8 and 1-of-16 multiplexers. Multiplexers in all three of these sizes are readily available in IC form.

**Table 10-1. This Truth Table
for a 1-of-4 Multiplexer Demonstrates That Only One Input Is Active at Any Time.**

Control Inputs		Main Inputs				Output
A	B	1	2	3	4	
0	0	0	x	x	x	0
0	0	1	x	x	x	1
0	1	x	0	x	x	0
0	1	x	1	x	x	1
1	0	x	x	0	x	0
1	0	x	x	1	x	1
1	1	x	x	x	0	0
1	1	x	x	x	1	1

(x = "don't care")

125

Table 10-2. Some Multiplexer Invert the Logic Signal Before the Output.

Control Inputs		Main Inputs				Output
A	B	1	2	3	4	
0	0	0	x	x	x	1
0	0	1	x	x	x	0
0	1	x	0	x	x	1
0	1	x	1	x	x	0
1	0	x	x	0	x	1
1	0	x	x	1	x	0
1	1	x	x	x	0	1
1	1	x	x	x	1	0

(x = "don't care")

Some multiplexers, it should be noted, invert the main input signals before feeding them to the output. In some cases this may be an advantage, in others it may turn out to be a disadvantage. Often it really won't make much difference. The truth table for an inverting 1-of-4 multiplexer is given in Table 10-2. This is the same as Table 10-1, of course, except the output states are reversed.

Multiplexers can take the place of complex gating circuits. For example, consider the truth table that is shown in Table 10-3. At best, it would be a definite nuisance to generate this truth table

Table 10-3. This Truth Table Would be Difficult to Implement Using Standard Gates.

Inputs				Output	$\overline{\text{Output}}$
A	B	C	D		
0	0	0	0	0	1
0	0	0	1	0	1
0	0	1	0	1	0
0	0	1	1	0	1
0	1	0	0	0	1
0	1	0	1	1	0
0	1	1	0	1	0
0	1	1	1	0	1
1	0	0	0	0	1
1	0	0	1	1	0
1	0	1	0	1	0
1	0	1	1	1	0
1	1	0	0	0	1
1	1	0	1	0	1
1	1	1	0	1	0
1	1	1	1	0	1

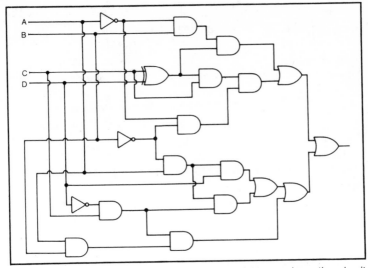

Fig. 10-2. The truth table from Table 10-3 requires a fairly complex gating circuit to implement.

using separate gates. A typical circuit for generating this truth table is shown in Fig. 10-2.

A 1-of-16 multiplexer can accomplish the same thing with a much simpler circuit. The 74150 is a typical 1-of-16 multiplexer in

Data input 7 ⊏	1	24 ⊐ Vcc
Data input 6 ⊏	2	23 ⊐ Data input 8
Data input 5 ⊏	3	22 ⊐ Data input 9
Data input 4 ⊏	4	21 ⊐ Data input 10
Data input 3 ⊏	5	20 ⊐ Data input 11
Data input 2 ⊏	6	19 ⊐ Data input 12
Data input 1 ⊏	7	18 ⊐ Data input 13
Data input 0 ⊏	8	17 ⊐ Data input 14
Enable ⊏	9	16 ⊐ Data input 15
Output ⊏	10	15 ⊐ Control input A
Control input 0 ⊏	11	14 ⊐ Control input B
Ground ⊏	12	13 ⊐ Control input C

Fig. 10-3. The 74150 is a 1-of-16 multiplexer in IC form.

IC form. This chip, which is part of the TTL family (see Chapter 12) is illustrated in the pinout diagram of Fig. 10-3.

Since there are 16 main, or data inputs, four control inputs (labeled A through D) are needed to identify each input address. These control inputs will correspond to our logic inputs in the truth table.

The 74150 happens to be an inverting type multiplexer, so we feed each main input with the opposite of the output logic state desired for the appropriate combination of control inputs. For example, when the control inputs are set to 1010, we want an output of logic 1, so we put the opposite value (logic 0) into input 10 (pin 21).

The complete circuit for generating the truth table of Table 10-3 using the 74150 1-of-16 decoder IC is shown in Fig. 10-4. Notice how much simpler and easier it is to follow this diagram compared to the one in Fig. 10-2.

Literally any truth table can be readily generated using a multiplexer in this manner. In some cases it may actually be more

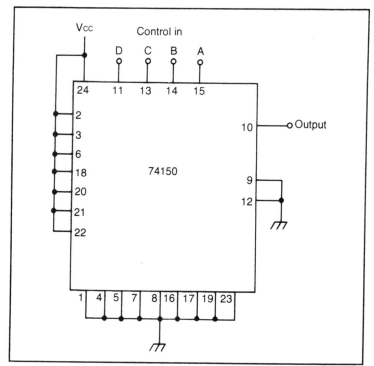

Fig. 10-4. The 74150 can be used to easily generate the truth table of Table 10-3.

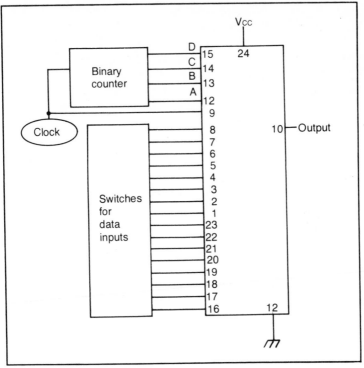

Fig. 10-5. Combining the 74150 1-of-16 multiplexer with a 4-stage, modulo-16 counter produces a binary pattern generator.

convenient to use separate standard gates, but a multiplexer can really come in handy when complex and/or unusual truth tables are called for.

A 1-of-16 multiplexer like the 74150 can generate 2^{16} different truth tables. In other words, there are 65,536 possible combinations of inputs and outputs.

A multiplexer can also come in handy when unusual counting sequences are called for. For instance, combining the circuit of Fig. 10-4 with a 4-stage, modulo-16 counter as shown in Fig. 10-5 would produce the following output pattern:

Clock Pulse #	Counter Outputs	Output
1	0001	0
2	0010	1
3	0011	0
4	0100	0

Clock Pulse #	Counter Outputs	Output
5	0101	1
6	0110	1
7	0111	0
8	1000	0
9	1001	1
10	1010	1
11	1011	1
12	1100	0
13	1101	0
14	1110	1
15	1111	0
16	0000	0
17	0001	0	(The pattern begins to repeat here.)
18	0010	1
19	0011	0
20	0100	0
21	0101	1
22	0110	1
23	0111	0
24	1000	0

and so on. Any pattern of binary digits (or bits) can be easily generated using this method.

Multiplexers have many additional applications. For example, they can be used to scan a series of switches or a keyboard, testing each switch to see if it is open or closed and telling the rest of the circuitry to respond accordingly.

DEMULTIPLEXERS

The opposite of a multiplexer is a demultiplexer. Where the control inputs on a multiplexer determine which of several inputs will be seen by the output, the control inputs on a demultiplexer determine which of several outputs will respond to the single input.

A simple demultiplexer circuit can be built from NAND gates, as shown in Fig. 10-6. The action of this circuit is demonstrated in the truth table given in Table 10-4.

Figure 10-7 shows the pinout diagram for a popular 1-of-16 demultiplexer IC. This is the 74154. Like the 74150 1-of-16 multiplexer described in the previous section, the 74154 belongs to the

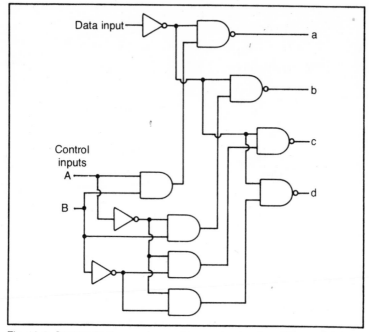

Fig. 10-6. Standard gates can be used to construct a demultiplexer circuit.

TTL family of digital devices (see Chapter 12).

A demultiplexer is often employed to convert binary numbers into another number system. For example, the 74154 can convert four digit binary numbers into single digit hexadecimal (base 16) numbers with the circuit illustrated in Fig. 10-8. For any combination of control inputs (any binary number from 0000 to 1111), one, and only one of the outputs will be at logic 0. The other fifteen outputs will be at logic 1 (Table 10-5).

Control Inputs	Main Input	Outputs			
A B		a	b	c	d
0 0	0	0	1	1	1
0 0	1	1	1	1	1
0 1	0	1	0	1	1
0 1	1	1	1	1	1
1 0	0	1	1	0	1
1 0	1	1	1	1	1
1 1	0	1	1	1	0
1 1	1	1	1	1	1

Table 10-4. A 1-of-4 Demultiplexer Has Only One Active Output at Any Time.

131

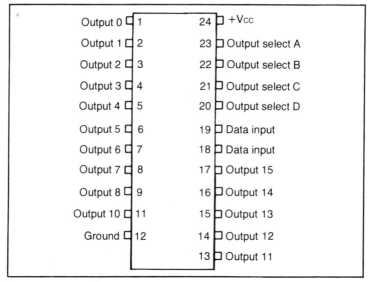

Fig. 10-7. The 74154 is a 1-of-16 demultiplexer in IC form.

The term demultiplexer is often abbreviated as DEMUX. Since these devices are frequently used for decoding data, they are also occasionally called decoders. By the same token, multiplexers

Table 10-5. This Truth Table Shows How a 1-of-16
Demultiplexer Can Be Used as a Binary to Hexadecimal Decoder (See Fig. 10-8).

Control Inputs				Outputs															
A	B	C	D	0	1	2	3	4	5	6	7	8	9	10	11	12	13	14	15
0	0	0	0	0	1	1	1	1	1	1	1	1	1	1	1	1	1	1	1
0	0	0	1	1	0	1	1	1	1	1	1	1	1	1	1	1	1	1	1
0	0	1	0	1	1	0	1	1	1	1	1	1	1	1	1	1	1	1	1
0	0	1	1	1	1	1	0	1	1	1	1	1	1	1	1	1	1	1	1
0	1	0	0	1	1	1	1	0	1	1	1	1	1	1	1	1	1	1	1
0	1	0	1	1	1	1	1	1	0	1	1	1	1	1	1	1	1	1	1
0	1	1	0	1	1	1	1	1	1	0	1	1	1	1	1	1	1	1	1
0	1	1	1	1	1	1	1	1	1	1	0	1	1	1	1	1	1	1	1
1	0	0	0	1	1	1	1	1	1	1	1	0	1	1	1	1	1	1	1
1	0	0	1	1	1	1	1	1	1	1	1	1	0	1	1	1	1	1	1
1	0	1	0	1	1	1	1	1	1	1	1	1	1	0	1	1	1	1	1
1	0	1	1	1	1	1	1	1	1	1	1	1	1	1	0	1	1	1	1
1	1	0	0	1	1	1	1	1	1	1	1	1	1	1	1	0	1	1	1
1	1	0	1	1	1	1	1	1	1	1	1	1	1	1	1	1	0	1	1
1	1	1	0	1	1	1	1	1	1	1	1	1	1	1	1	1	1	0	1
1	1	1	1	1	1	1	1	1	1	1	1	1	1	1	1	1	1	1	0

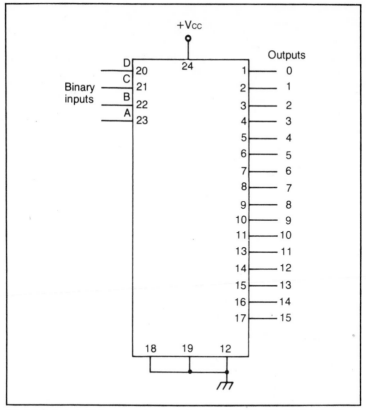

Fig. 10-8. A demultiplexer can be used as a binary to hexadecimal converter.

may be referred to as encoders.

If the four control inputs are fed with the outputs of a binary clock, the outputs will cycle through in sequence, like this:

0 - 1 - 2 - 3 - 4 - 5 - 6 - 7 - 8 - 9 - 10 - 11 - 12 - 13 - 14 - 15 - 0 - 1 - 2 - 3 - 4 - 5 - 6 - 7 - 8 - 9 - 10 - 11 - 12 - 13 - 14 - 15 - 0 - 1 - 2 - 3 - and so on.

For even more versatility, most binary clocks have a CLEAR input which sets the clock back to 0000 and starts over. By feeding back the appropriate output from the demultiplexer to the CLEAR input of the counter, the counting sequence can be limited to any maximum value. For instance, feeding back demultiplexer output 8 to the counter CLEAR input, will produce the following output sequence:

0 - 1 - 2 - 3 - 4 - 5 - 6 - 7 - 8 - 0 - 1 - 2 - 3 - 4 - 5 - 6 - 7 - 8 - 0 - 1
- 2 - 3 - and so on.

Note that with some circuits an inverter may have to be used on this feedback signal for correct operation. If the wrong logic signal is fed into the CLEAR input, the output will be stuck in a rather nonuseful rut:

0 - 0 - 0 - 0 - 0 - and so on.

DISPLAY DRIVERS

Many digital circuits use seven-segment LED or LCD displays as output devices, as described in Chapter 8. Obviously, some method of converting the binary signals into the appropriate lit displays is necessary. Individual gates may be used, but this tends to increase total circuit bulk and cost. Since this is such a common requirement in digital electronics, a number of display driver ICs have been made available.

A typical example is the CD4511 BCD-TO-7-Segment Latch/Decoder/Driver IC, which is illustrated in Fig. 10-9. As the rather lengthy name indicates, this chip fills a number of related functions. Basically it accepts a four digit BCD (Binary Coded Decimal) number (from 0000 to 1001) and puts an output signal on the appropriate pins to light the desired display segments.

BILATERAL SWITCHES

It would often be handy to have a digitally controlled switch.

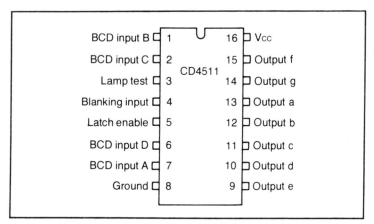

Fig. 10-9. The CD4511 is a BCD to 7-segment latch/decoder/driver in IC form.

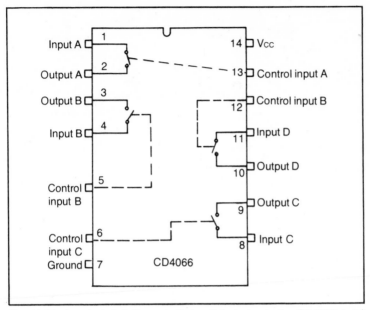

Fig. 10-10. A typical digitally controlled switch device is the CD4066 quad bilateral switch.

That is, a switch that may be open or shut depending on the logic signal from another part of the circuit. This function can usually be served with a gating network, but this can often be awkward, bulky, and expensive. As you've probably anticipated, integrated circuit manufacturers have met this need with specialized chips.

Figure 10-10 shows the pinout diagram and functional internal structure of the CD4066 quad bilateral switch. This CMOS chip (see Chapter 12) consists of four digitally controlled switches. The switches are called bilateral because they have no fixed polarity. The pins labeled as switch inputs on the diagram may be used as switch outputs, and vice versa. Which end of the switch is the input and which is the output does not matter.

Digital switch units like the CD4066 are often used in hybrid circuits that use both digital and analog devices. Analog components like resistors or capacitors may be selected or programmed via digital signals. Digital to analog signal conversion is one obvious application for this type of device.

TRI-STATE LOGIC

A somewhat similar concept is that of tri-state logic, although this technique is used only for switching digital signals. It is inap-

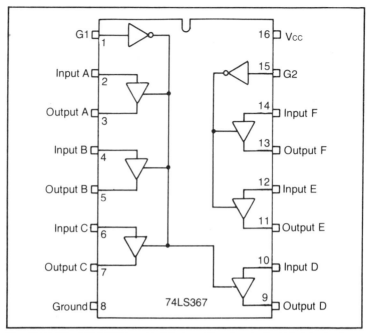

Fig. 10-11. The 74LS367 hex tri-state bus driver is a typical tri-state device.

propriate for analog applications. Most digital circuits have two possible output conditions—either the output is a low voltage (logic 0) or it is a high voltage (logic 1). Tri-state systems add a third possible output state. This is a high impedance state, which is neither a logic 0 or a logic 1, and appears as the absence of any signal at all.

Figure 10-11 shows a typical tri-state device. This is the 74LS367 hex tri-state bus-driver chip. If the appropriate G input is at logic 0, the buffers behave just like ordinary buffers—the output is the same as the input. However, when a buffer's G input goes high, that buffer is effectively cutoff, and the output goes to the third

**Table 10-6. This Truth Table for a Tri-State
Buffer Demonstrates the Principle of Tri-State Logic.**

G Input	Data Input	Output
0	0	0
0	1	1
1	x	high-impedance

(x = "don't care")

136

high-impedance state. The truth table for this device is shown in Table 10-6.

AND STILL OTHER DEVICES

Many other digital devices also exist, and more are being developed every month. In this book so far we have covered the more important and most commonly encountered devices. Many of the digital devices and circuits we have not explored here are fairly specialized and have relatively limited applications. This is especially true of LSI (Large-Scale Integration) ICs, such as microprocessors and digital voltmeters. Since complex digital circuits are made up of simpler "building block" subcircuits, SSI (Small-Scale Integration) devices have a much wider range of potential applications, since they can be combined to form the circuits contained in MSI (Medium-Scale Integration) and LSI chips. The basic gates described in Chapters 2, 3, and 4 are the lowest level of digital circuitry.

One group of LSI devices deserves special mention. These are digital memory circuits, which have many important applications and will be discussed in the next chapter. The following two chapters will discuss the vital concepts of logic families (Chapter 12) and negative logic (Chapter 13). After that we will start to put our knowledge of digital logic components to work in several practical projects.

Chapter 11

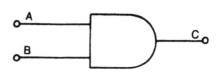

Memory Circuits

In many digital electronics applications, it is essential to have some way to store digital data (binary numbers). This requirement is particularly pressing with digital systems like computers and related equipment. If the amount of data to be stored is relatively small, the task can be accomplished with some of the digital devices discussed in the previous chapters. A flip-flop, for example, can store a single bit (binary digit). A binary word (a string of binary digits of a specific length) can be stored in a shift register. Unfortunately, these techniques are extremely limited. Unless the amount of data to be stored is quite small, flip-flops and shift registers are likely to prove totally inadequate for the task. Fortunately, specialized digital circuits called memories have been developed, and many different types and sizes of digital memories are available in integrated circuit form.

RAM

One of the basic types of digital memory is called RAM. This is an abbreviation for *random access memory*. What this name means is that any specific location in the memory can be contacted without stepping through any of the other locations.

A shift register, on the other hand, is an example of a sequential memory. For instance, let's say we need to know the logic state of bit number 5 in an eight-stage SISO shift register. Before we can reach bit 5, we have to step through bits 1, 2, 3, and 4. Obviously, this will take some finite amount of time. When dealing with thousands of bits of data, the disadvantages of a sequential memory become painfully obvious.

A random access memory, however, assigns a unique address to each individual storage location, with each location holding one piece of data (either a bit, or a word). This type of memory can be considered as an analogy to a post-office box system. Any box (or memory location) can uniquely be defined by an address identifying its column and row. This idea is illustrated in Fig. 11-1.

Any specific piece of data can easily be found by going directly to its unique address. Rather than having to look at each storage location in sequence (1 - 2 - 3 - 4 - 5 - 6 - 7 - etc.), we can go straight to the desired address. For example, at location C7, we might find 0101.

In a RAM we can either look at the value stored at a given address without changing it, or we can replace the old value with a new value. We can either read old data, or write new data. For this reason, RAM is sometimes called read/write memory, or RWM. Some technicians feel this is a better name (since the ROM, or *read only memory* discussed in the next section, can also be randomly accessed), but RAM is the established name in common usage. There are two basic types of RAM. They are static RAM, and dynamic RAM.

Static RAM

A static RAM is basically made up of a series of addressable flip-flops. Data can be stored in a static RAM virtually indefinitely, unless the stored values are erased, or the power supply is interrupted. A static RAM can not store data without continuously

	A	B	C	D
1	A1	B1	C1	D1
2	A2	B2	C2	D2
3	A3	B3	C3	D3
4	A4	B4	C4	D4
5	A5	B5	C5	D5

Fig. 11-1. Memory addressing works something like a post office box system.

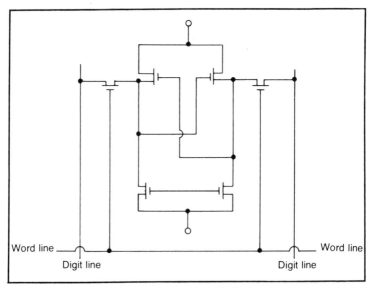

Fig. 11-2. A static memory cell is basically a flip-flop.

applied power. A typical static memory cell using CMOS technology (see Chapter 12) is illustrated in Fig. 11-2.

One of the earliest static RAM ICs was Intel's MM1101, which first was available in the late 1960s. The pinout diagram for this chip is shown in Fig. 11-3. Eight pins are used for defining the memory address (A0 through A7). This means that the memory addresses may range from binary 0000 0000 (decimal 0) to 1111 1111 (decimal 255). In other words, the MM1101 static RAM IC contains 256

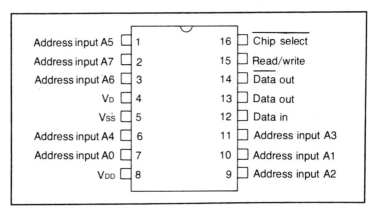

Fig. 11-3. An early static RAM chip was the MM1101 256-bit static RAM IC by Intel.

140

independent memory locations, each of which can store the value of a single bit (binary digit—0 or 1).

Larger static RAM ICs have been developed since then. In larger memories, each memory address will store more than one digit. For example, a 256 × 4 RAM would store 1024 bits in the form of 256 independent 4-bit words.

Dynamic RAM

The other type of RAM is dynamic RAM. In this case, each bit is stored in a capacitor. A charged capacitor represents a logic 1, while a discharged capacitor would stand for a logic 0.

A dynamic memory cell is much simpler than a static memory cell. Compare the dynamic memory cell in Fig. 11-4 with the static memory cell that was shown in Fig. 11-2. This means that a dynamic memory of a given storage capability will tend to be much smaller and less expensive than a comparable static memory.

The MM5270 dynamic RAM IC, which is shown in Fig. 11-5 can store 4096 independent bits of data. The 12 address lines allow for values ranging from 0000 0000 0000 (decimal 0) to 1111 1111 1111 (decimal 4095).

However, dynamic RAM is not without its disadvantages. The most important of these is that no capacitor can hold a charge indefinitely. Eventually the charge will tend to leak off.

Electronically reading the value stored in a dynamic memory cell tends to recharge partially charged capacitors, refreshing the memory. Practical dynamic memory systems, therefore, require

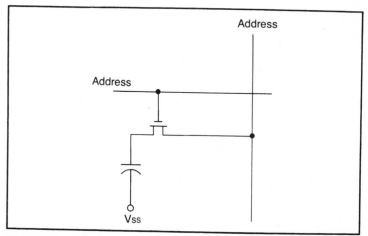

Fig. 11-4. In a dynamic memory cell, each bit is stored in a capacitor.

141

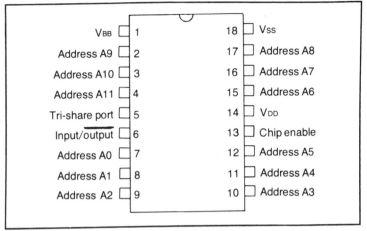

Fig. 11-5. The MM5270 is a 4096-bit dynamic RAM chip.

refreshing circuits that will automatically read all of the memory locations at regular intervals to prevent the charged capacitors from leaking off too much voltage.

A dynamic memory cell is simpler than a static memory cell, but dynamic memories require more complex supporting circuitry (to periodically refresh the capacitor charges), so a trade-off is inevitable.

Improvements in technology have allowed for more static memory cells to be contained in a single IC chip, and at somewhat lower manufacturing costs, so at the moment, the balance of the scales tips somewhat towards static over dynamic memories in most general applications.

ROM

Data can be read out of or written into a RAM. The user can store his own data freely changing or updating it at any time. However, if the power to the system is cut off for any reason, any data stored in a RAM will be irretrievably lost. In some applications, there may also be data to be stored in the memory in a way that the user can not inadvertently erase or change it. The solution is a memory that can be read from, but not written to. This type of memory is called a ROM, or read only memory.

All of the data in a ROM is permanently determined by the manufacturer when the chip is made. Obviously, ROMs are only practical for applications where many identical units are required. For example, a computer manufacturer might include instructions for the machine to understand BASIC (a programming language) so that the user won't have to load this information into the computer each time he turns it on.

Since the data stored in a ROM is permanently hard-wired within a chip, each ROM memory cell can be much simpler than either static or dynamic RAM cells. Several ROM memory cells are shown in Fig. 11-6. The data stored in each cell is determined by the presence (logic 1) or the absence (logic 0) of a connecting diode at the appropriate address location. The user has no way of changing any of the data stored in a ROM. If even a single bit must be changed, the entire ROM chip must be replaced.

PROM

Since the data in a ROM must be determined at the time of manufacture, this type of memory is not very practical for applications where only a few copies are to be made. Unless you need a few hundred identically programmed ROM chips, the cost per unit would be too high.

For applications where the permanent storage capabilities are needed, but only one or a few units will be needed, a user programmable ROM has been developed. This type of memory is called a *programmable read only memory*, or PROM. Each memory cell in a PROM is similar to a ROM cell. The diodes are a special fused type.

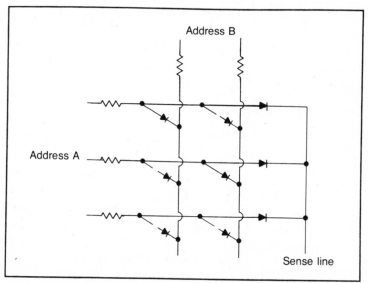

Fig. 11-6. ROM memory cells are much simpler than RAM memory cells.

143

Every memory cell in a PROM contains a fused diode at each and every address location. The user can program the chip by blowing the fuses on unwanted diodes.

Once programmed, a PROM behaves exactly like a ROM. Data can be read from it, but the stored data cannot be erased or changed. Actually, more diodes can be blown (1's changed to 0's), but there is no way to replace a blown diode (change a 0 to a 1).

To blow a fused diode in a PROM a slightly higher than normal voltage is needed, so that a special programming circuit may be used. If a PROM is used in a computer, for example, you don't need to worry about accidentally blowing additional diodes.

Once programmed, the data stored in a PROM cannot be changed (except as noted above). This means if a mistake is made, or even a small change is needed, the PROM must be discarded and a new one must be programmed.

EPROM

A special type of PROM allows the chip to be reused. Data can be erased and new data can be written. Not surprisingly, this type of memory is called an *erasable programmable read only memory*, or EPROM.

An EPROM works in essentially the same way as a regular PROM, except for the fact that the entire chip can be cleared (all of the stored data is erased) by exposing the chip to a strong ultraviolet light source. There is no way to change just a few bits. It is an all or nothing proposition.

Since sunlight and other visible light sources contain some ultraviolet energy, a programmed EPROM should be shielded from all light to avoid accidental erasure of data.

MEASURING MEMORY SIZE

Since binary numbers are used to define the memory location addresses, the number of cells in a memory system is almost invariable a power of 2, like 256 (2^8), 1024 (2^{10}), or 4096 (2^{12}). In large practical systems, memory size is usually defined as being so many K. K is normally used to indicate a factor of one thousand. However, 1000 is not a power of 2. The nearest power of 2 is 1024, so in memory systems, K actually represents a factor of 1024. 1 K = 1024, 4 K = 4096, 16 K = 16,384, and 64 K = 65,536. These are the most common memory sizes in small computers.

You do have to be careful about just what is being counted. For memory ICs, 1 K indicates a storage capability of 1024 bits, while in computer systems, the quantity in question is the number of bytes (or 8-bit binary words) that can be stored. A 4 K memory in this instance would hold 4096 sets of 8 bits each, or 32,768 individual bits (0's or 1's). The terminology can cause some confusion if you're not careful.

Chapter 12

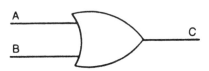

Logic Families

In digital electronics there are a number of ways to accomplish the same results. A number of technologies for forming gates on IC chips have been developed. The primary difference between the various technologies lies in the basic circuits that are used. ICs designed according to a specific technology are said to belong to the same logic family. In this chapter we will examine a few of the most important of the common logic families.

RTL

Most of the earliest digital integrated circuits belonged to the RTL family. RTL is short for *resistor-transistor logic*. As the name suggests, the basic circuit is made up of a resistor and a transistor.

Figure 12-1 shows the most basic RTL circuit. This is a simple inverter. A RTL NOR gate circuit is shown in Fig. 12-2. The base resistor in each case is used to limit the input current.

Most RTL chips were designed for a power supply of 3.6 volts. RTL circuits are fairly easy to implement in chip form, so they were popular in the early days. However, faster and more powerful technologies have since been developed at competitive costs, so today RTL is pretty much an obsolete logic family.

DTL

Another popular early logic family was DTL, or *diode-transistor logic*. Basically, DTL gates are similar to RTL circuits

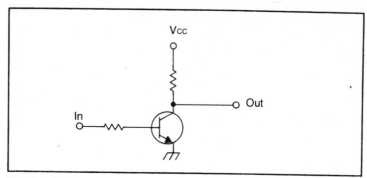

Fig. 12-1. A RTL inverter circuit is made up of an input resistor and a transistor.

with the addition of input diodes. A DTL NOR gate is shown in Fig. 12-3. Compare this with the RTL NOR gate of Fig. 12-2. Due to its relatively slow switching speed, DTL, like RTL, is more or less an obsolete technology.

TTL

Digital electronics didn't really begin to take off in a big way until the development of the TTL logic family. TTL stands for *transistor-transistor logic*. This logic family is based upon the ability to form transistors with multiple emitters (see Fig. 12-4) on an integrated circuit chip. TTL gates offer relatively high speed operation, low power consumption, and a reasonably small susceptibility to noise.

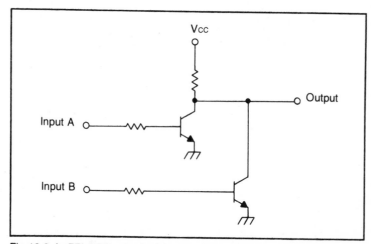

Fig. 12-2. An RTL NOR gate circuit is essentially an extension of the RTL inverter shown in Fig. 12-1.

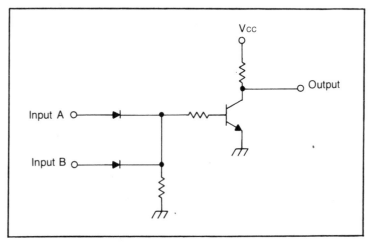

Fig. 12-3. A DTL NOR gate uses input diodes instead of input resistors.

A typical TTL NAND gate circuit is illustrated in Fig. 12-5. The multiple emitters essentially take the place of the diodes in DTL gates.

The power supply voltage for TTL devices is fairly critical. These ICs require a tightly regulated 5-volt power supply. Generally power supply deviations greater than ± 0.5 volt cannot be tolerated. An incorrect supply voltage will almost certainly result in improper operation, and is quite likely to actually damage the ICs (especially over-voltages).

An input voltage below 0.2 volt is interpreted by TTL gates as a logic 0. A logic 1 is defined as anything above 2.4 volt. An input voltage between 0.2 volt and 2.4 volt is undefined in TTL logic. This wide gap provides a comfortable noise margin. Figure 12-6

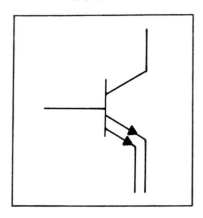

Fig. 12-4. TTL circuits are built around multi-emitter transistors.

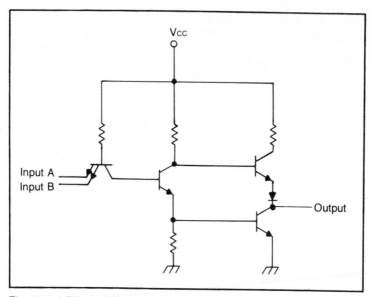

Fig. 12-5. A TTL NAND gate is more complex than RTL or DTL circuits.

shows how this ignored in-between range allows the gates to be uninfluenced by noise on the input lines, even if the noise is rather severe, as illustrated here.

A logic 1 output from a TTL gate will not be much more than about half of the power supply voltage (nominally 2.5 volts). A logic 0 input will require the input device to sink a fairly substantial amount of current. 16 mA is a typical figure.

If an input to a TTL device is left unconnected (or floating), it will behave like a logic 1 input. A floating input pulls itself up to the

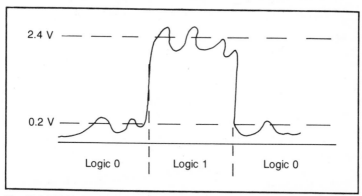

Fig. 12-6. TTL's wide noise margin allows circuits to ignore noise pulses between 0.2 and 2.4 volts.

logic 1 voltage level. Most TTL gates have a fan-in of one, and a fan-out of about ten (see Chapter 2).

A TTL logic 1 output voltage can be raised with a pull-up resistor, as shown in Fig. 12-7. This trick is frequently used by digital circuit designers.

A large number of TTL devices are readily available in IC form. Generally they follow the 74xx numbering scheme (7400, 7401, 7403 . . . 74154, 74155, and so forth). You may also encounter TTL ICs numbered 54xx. These devices are essentially identical to the 74xx series (5400 = 7400, 5401 = 7401, and so on), except they are designed to meet stricter military specification (primarily, a wider range of operating temperatures). For the vast majority of users, there is no functional difference between 54xx and 74xx series devices.

The TTL logic family enjoys a well deserved popularity. There are a number of significant advantages to this technology. For one thing, its very popularity has ensured the wide-spread availability of literally hundreds of SSI and MSI devices at reasonably low prices. TTL devices are relatively immune to noise, and consume fairly small amounts of power during steady state (the output does not change states) conditions.

Perhaps the most important of the advantages of TTL is the circuitry's capability for operating at high speeds. Most TTL devices can handle 20 MHz (20,000,000 hertz) signals without problems. A few special devices can even accept signals ranging up to 125 MHz (125,000,000 hertz).

Unfortunately, no technology is perfect, and TTL is no exception to that rule. There are some disadvantages. First, while TTL is fine for SSI (Small-Scale Integration) and MSI (Medium-Scale Integration) ICs, the technology does not appear to be particularly

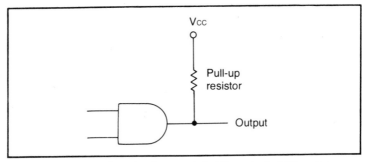

Fig. 12-7. A pull-up resistor is often needed at the output of TTL gates.

well suited to LSI (Large-Scale Integration) devices.

Probably the biggest disadvantage of TTL circuitry is power consumption. We have already mentioned that TTL gates don't dissipate too much current for steady-state (constant output) conditions. However, during the instant of output switching (changing from a 0 to a 1, or vice versa) a large current spike is pulled from the power supply for a fraction of a second. This may be insignificant in small circuits with just a handful of gates, but in large systems with dozens, or even hundreds of gates, the power drain might be quite severe.

In addition, this current spike can be dangerous to the delicate circuitry within the IC itself. For this reason a bypass capacitor should be placed across the power lines on each and every TTL IC in every circuit. A 0.001 μF to 0.01 μF capacitor connected between the ground and V_{cc} pins will usually be sufficient. This capacitor should be mounted as physically close to the IC itself as possible. Every TTL IC should have its own individual bypass capacitor. This is especially crucial in large systems where the cumulative current spike could be extremely large.

Actually, TTL is not just a single logic family. A number of subfamilies have been developed, each with their own special advantages.

Low-Power TTL

While a typical TTL gate doesn't consume very much current by itself, large systems with many TTL devices can end up drawing a considerable amount of power. The low-power TTL subfamily was designed to ease the power requirements. A typical low-power TTL gate uses about 0.1 of the current consumed by its regular TTL equivalent. Of course, you don't get something for nothing. A trade-off is inevitable. In this case, the choice is between power consumption and operating speed. Low-power TTL devices can only function about 0.1 to 0.25 as fast as their regular TTL equivalents.

Low-power TTL ICs are generally interchangeable with regular TTL units, and are numbered in the same way. An L is added to the middle of the number to identify the device as a low-power type. The numbering takes the form of 74Lxx. For example, a 74L04 is the low power equivalent for a 7404. Low-power TTL can be a good choice for circuits requiring a fairly large number of ICs, but where operation at high frequencies is not demanded.

High-Speed TTL

In some applications, high operating speed may be more important than power consumption. When regular TTL isn't fast enough, the compromise can be taken to the other extreme with the high-speed subfamily of TTL.

High-speed TTL devices consume twice the power of regular TTL chips, and can operate at twice the speed. Some special purpose units, such as counters, may be able to handle frequencies up to 50 MHz (50,000,000 Hz).

Gate inputs in high-speed ICs draw more current than regular TTL, so the fan-in for this subfamily is 1.3. That is, each gate looks like one and a third gates to the preceding output. The fan-out for high-speed TTL remains at about 10. An output with a fan-out of 10 can drive about seven high-speed TTL inputs.

High-speed TTL devices are identified with an H. A 74H75, for example, is the high-speed equivalent of a regular TTL 74L75, or a low-power TTL 74L75. The custom of identifying the subfamily with a mid-number letter is followed for most variations on standard TTL chips.

Schottky TTL

As integrated circuit technology improved over the years, a somewhat more advanced form of TTL was developed. Very fast diodes (known as Schottky diodes) are included on these chips to create a better speed/power trade-off than other types of TTL. A typical Schottky TTL circuit uses about as much power as an equivalent high-speed TTL device, but is capable of almost twice as much speed. Schottky TTL units can be operated at frequencies 3.5 times faster than standard TTL. Schottky TTL devices are identified with an S in the part number, such as 74S32 (the Schottky equivalent for a standard TTL 7432).

A low-power Schottky TTL subfamily also exists. These units run at about the same speed as standard TTL, but consume only

Table 12-1. Comparing the Operating Speed and Power Consumption of TTL and its Subfamilies Illustrates the Speed/Power Trade-Off.

	Standard TTL	Low-power TTL	High-speed TTL	Schottky TTL	Low-power Schottky TTL
speed	1	0.1	2	3.5	1
power	1	0.1	2	2	0.2

about 20% of the power. Low-power Schottky ICs are marked LS. For instance, 74LS32. The various subfamilies of TTL are compared in Table 12-1.

There is one more recently developed TTL subfamily that is likely to have growing importance in digital electronics, but before we deal with this one, we need to look at another major logic family known as CMOS.

CMOS

While TTL (and its various subfamilies) continues to hold an extremely important place in digital electronics, it has been over-shadowed to a large degree in recent years by another logic family called CMOS (for *complementary metal-oxide semiconductor*). Occasionally this family may be called COS/MOS.

CMOS ICs are made from P-channel and N-channel metal-oxide semiconductor transistors. This technology offers very low-power consumption and the ability to operate from a wide range of supply voltages (typically about 3 to 15 volts), at a relatively low cost.

Part of the price, however, is a trade-off in speed. The higher the power supply voltage (as long as it doesn't exceed the maximum, of course), the faster CMOS gates can operate. But they are still slower than equivalent TTL circuits.

CMOS circuitry is well suited for LSI (Large-Scale Integration) devices, as well as SSI (Small-Scale Integration) and MSI (Medium-Scale Integration) units. Most CMOS ICs offer a guaranteed noise immunity of $\pm$ 3 volts, and typical units exceed this specification. Practical noise immunity of $\pm$ 4.5 volt is not uncommon for a power supply of 10 volts. (Lower supply voltages lead to narrower noise margins.)

All inputs to CMOS gates must be connected to something. If they are not being used with a digital signal source, they should be tied to either the positive power supply, or ground. A floating input can make a CMOS gate quite unstable, and the output state may become unpredictable.

Many CMOS devices are of the tri-state type. In addition to the usual logic 0, and logic 1 output states, a tri-state circuit might also have a special high-impedance output state, which acts as an open circuit. That is, a CMOS device may be electrically switched off, and made to seem like it does not exist within the circuit.

A number of numbering schemes are in use for CMOS integrated circuits. The two most common are CD4xxx, and 74Cxx.

The 74Cxx system follows TTL numbering, and similar numbered devices (74C93, 7493, 74H93, and 74LS93, for instance) are functionally identical, and have the same pinout. This is for convenience. The 74Cxx is not a TTL subfamily, and should not be confused with TTL. The CD4xxx numbering system is unique to CMOS. Many 74Cxx and CD4xxx devices are interchangeable, but the numbers may not be similar. The arrangement of the pin connections may also be different. As an example, Fig. 12-8 shows the pinout diagram for a 74C04 and Fig. 12-9 is the pinout diagram for a CD4009. Both of these chips are hex inverters, and function in the same way, even though they are packaged and numbered differently.

One potential problem with CMOS ICs is that they tend to be sensitive to static electricity. Stored CMOS chips should have their pins shorted together to prevent accidental static discharge, which could damage the sensitive on-chip components. This can be done by inserting the pins in a special conductive foam, or storing the ICs in anti-static plastic containers. Ordinary aluminum foil also works fine.

Be careful not to touch the pins of a CMOS IC unless you are

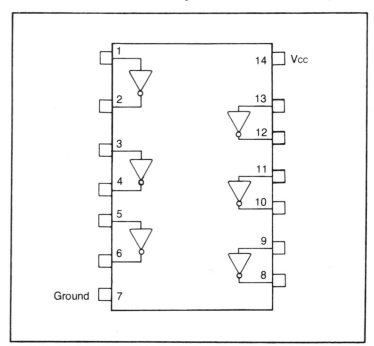

Fig. 12-8. The 74C04 hex inverter is a typical CMOS IC.

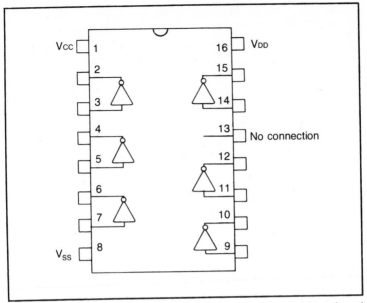

Fig. 12-9. The CD4009 hex inverter is similar to the 74C04, but it is packaged differently.

securely grounded. We've all had a spark jump from our fingers to a metal doorknob, or something similar. Imagine what such a burst of static voltage could do to the thin metal-oxide film in a CMOS chip!

If you solder the pins of a CMOS IC directly (rather than using a socket), your soldering iron should also be properly grounded for insurance against static electricity mishaps. Damage from static discharge was a severe problem with early CMOS devices. Most modern units contain on-chip protection which can help to ward off trouble. Even so, precautions are more than advisable when working with CMOS integrated circuits.

A TTL-CMOS HYBRID

A recent development has been a new hybrid logic family known as *high-speed CMOS*. The 74HCxx numbering scheme is followed for these devices. The high-speed CMOS family offers many of the advantages of both the ordinary CMOS and TTL families. High-speed CMOS devices require a just slightly higher supply current than their regular CMOS equivalents, but the power drain is still lower than for Schottky TTL ICs at low frequencies. In terms of speed, the high-speed CMOS family offers approximately a ten-fold increase in performance over standard CMOS circuits.

155

Operating speed capabilities for this logic family are roughly similar to low-power Schottky TTL.

These chips work best with a power supply in the range of 3 to 6 volts, but they are capable of operating from anything between —0.5 volt and 7.0 volts. Typical devices can dissipate a constant 500 milliwatts at room temperature. Most high-speed CMOS devices are pinout and functionally equivalent to 74LSxx low-power Schottky TTL ICs. The 74HCxx numbering system is used. In a few cases, where there is no equivalent in the 74xx numbering system, high-speed CMOS ICs are pinout and functionally compatible to devices in the CD4xxx numbering system.

Special devices are available to make interfacing high-speed logic chips to other logic families easy. These units are numbered 74HC4301 through 74HC4306. Several devices in this logic family will directly accept the logic voltage levels employed by low-power Schottky TTL.

ECL

Another important logic family is ECL, or *emitter-coupled logic*. ECL devices are not in as wide-spread use as TTL and CMOS units, but they are invaluable in applications where very high operating frequencies are called for. Some ECL devices can handle frequencies up to 200 MHz (200,000,000 Hz).

ECL integrated circuits also put an almost constant drain on the power supply. There are no steep current spikes when the outputs switch logic states (change from 0 to 1, or vice versa) as with the other major logic families. ECL devices are numbered 10xxx.

COMBINING LOGIC FAMILIES

It is often necessary to mix devices from different logic families with a single circuit. Some unusual devices may be available in one logic family. Often different portions of a large system may have different requirements. For example, in a circuit using many ICs, keeping the power consumption down is likely to be a major design consideration. However, a few stages within the system may need to be operated at high speeds. It would probably be wasteful and needlessly expensive to use high-speed ICs throughout the circuit if the high frequencies will only be encountered by a few of the devices. The solution is to mix logic families.

Unfortunately, different logic families cannot just be directly wired to one another willy-nilly. Different power-supply voltages

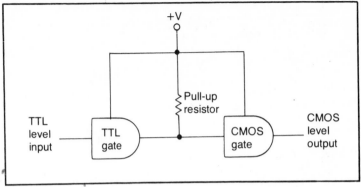

Fig. 12-10. A pull-up resistor can allow a TTL gate to drive a CMOS device.

are used and the logic levels are defined at different voltage levels. Input and output currents may also be mismatched, creating potential fan-in/fan-out problems. By using a few basic tricks, the various logic families and subfamilies can be successfully used together within a circuit.

The addition of a simple pull-up resistor, as illustrated in Fig. 12-10 allows TTL gates to drive CMOS devices without much trouble. The supply voltage should be suitable for the TTL ICs, that is, nominally 5 volts. CMOS devices will work with this lower voltage, although optimum performance generally requires a higher supply voltage. The value of the pull-up resistor will usually be fairly low. 1 K (1000 ohms) is typically used.

Some CMOS devices can drive TTL gates directly, as shown in Fig. 12-11. The power supply must be suitable for TTL use (5 volts). However, a CMOS driving TTL situation usually involve problems stemming from the ability of the CMOS device to supply

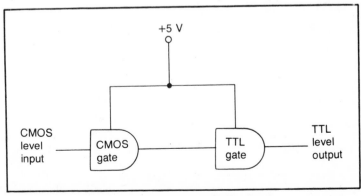

Fig. 12-11. Some CMOS circuits can drive TTL ICs directly.

157

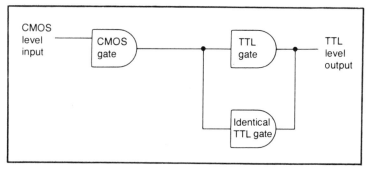

Fig. 12-12. Paralleling gates allows CMOS gates to drive TTL units.

and sink the required currents for each of the logic states. Typical TTL specifications call for current no more than -1.6 mA for a logic 0, and no more than 40 μA for a logic 1. In cases where a single device doesn't have sufficient current handling capability, two or more identical devices can be paralleled, as illustrated in Fig. 12-12.

Some CMOS devices are specially designed to make interfacing with TTL components easier. The CD4009A hex inverter and the CD4010A hex buffer are ideal for use as logic-level shifters. One terminal can take a high voltage (12 to 15 volts) from the CMOS portion of the circuit, while the 5 volts required for TTL can be fed into the same chip at a second terminal. Figure 12-13 demonstrates the use of a CD4010A buffer in interfacing a CMOS gate with a TTL device.

Connecting CMOS and ECL ICs can be a little trickier. ECL gates tend to work best with voltages between ground and -5

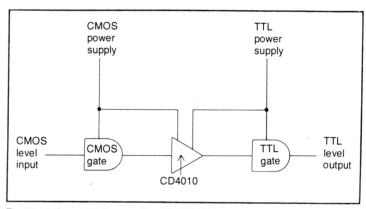

Fig. 12-13. Special buffers, such as the CD4010, can be used to interface CMOS and TTL chips.

158

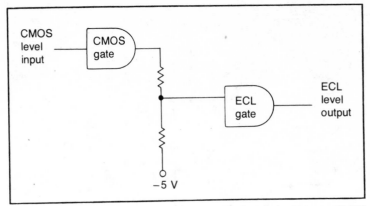

Fig. 12-14. A simple voltage divider network permits interfacing between CMOS and ECL devices.

volts. CMOS can be made to work with this voltage by using a voltage divider resistance network, as shown in Fig. 12-14.

SELECTING THE BEST LOGIC FAMILY

For many applications, the decision of which logic family to use won't really matter very much. Still, in a number of cases the trade-offs between cost, power consumption, and operating speed will be of vital importance. The choice can be compounded by the fact that not all devices are available in all families.

The majority of experimenters today seem to prefer CMOS as the best all around logic family for general purpose work, at least where very high frequencies are not involved. TTL runs a close second, especially in relatively small scale projects. There is no one best logic family. The decision must be based on availability and the requirements of the individual circuit in question.

To give you some experience with both of the major logic families, the projects presented in later chapters will employ both TTL and CMOS ICs. Generally, the decision of which family to use in these simple projects was arbitrary, and other logic families may be used if you prefer. If you decide to use a logic family other than the one presented for a given project, be careful that you use the appropriate power supply and follow the right specifications throughout the circuit.

Chapter 13

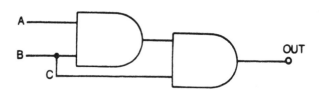

Negative Logic

Throughout most of this book positive logic is employed. This is the norm in digital electronics. However, sometimes, using negative logic can be a useful technique for the digital circuit designer. In positive logic, a logic 1 is represented by a higher voltage than a logic 0. In negative logic, everything is reversed. The higher voltage represents a logic 0, and the near ground signal state stands for a logic 1.

So what's the difference between positive and negative logic? In functional terms, there is no difference. The two forms of logic are simply two different ways of thinking about the action of digital gates. For example, consider the truth table for an AND gate, which is shown in Table 13-1. This truth table is in positive logic form. The 1's are high, and the 0's are low.

Now compare Table 13-1 with Table 13-2, which shows the negative logic truth table for the same gate. Now, the 1's are low, and the 0's are high. Every input and output signal reverses its nominal logic state, even though the actual voltages in the circuit remain exactly the same. The only thing that has changed is the names we choose to identify each of the possible input and output conditions.

Look at the negative logic AND gate truth table very closely. It should look familiar. The output is a logic 0 if, and only if both inputs are at logic 0, or, in other words, if either input A or input B is a logic 1, then the output will be a logic 1. What behaves as an AND gate in

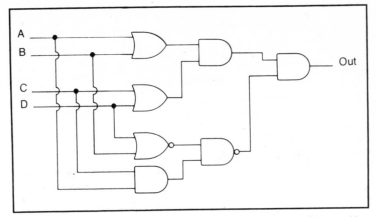

Fig. 13-1. Possible solution to the example described in the text from positive logic.

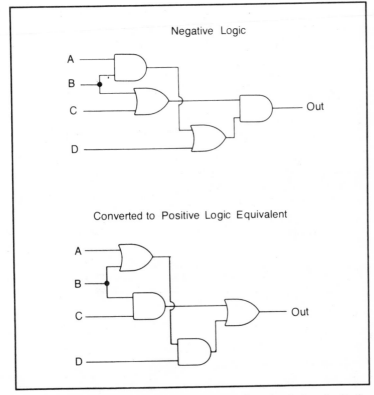

Fig. 13-2. Using negative logic, a simpler solution to the example described in the text can be found.

Inputs		Output
A	B	
0	0	0
0	1	0
1	0	0
1	1	1

Table 13-1. The Truth Table for
an AND Gate Using Positive Logic.

positive logic, functions as an OR gate in negative logic. It is absolutely essential to remember that the actual circuitry performs in precisely the same way in either case. The only thing we have altered is our way of interpreting the action of the circuit.

Similarly, Table 13-3 compares the positive and negative logic truth tables for a (positive logic) NAND gate. Notice how a positive logic NAND gate also functions as a negative logic NOR gate. As

Table 13-2. When
Using Negative Logic in
a Truth Table for an AND Gate.
All Inputs and Outputs Are Inverted.

Inputs		Output
A	B	
1	1	1
1	0	1
0	1	1
0	0	0

you might suspect, a positive logic OR gate behaves as a negative logic AND gate. This is demonstrated in Table 13-4.

The positive and negative logic truth tables for a (positive logic) NOR gate are given in Table 13-5. The positive logic NOR gate is the equivalent of a negative logic NAND gate.

In the case of an Exclusive-OR (X-OR) gate, the negative logic

Table 13-3. As These Truth Tables Show, a
Positive Logic NAND Gate Is a Negative Logic NOR Gate.

Positive			Negative		
Inputs		Output	Inputs		Output
A	B		A	B	
0	0	1	1	1	0
0	1	1	1	0	0
1	0	1	0	1	0
1	1	0	0	0	1

Table 13-4. Here the Positive and Negative Logic Truth Tables for an OR Gate Are Compared.

Positive		Negative	
Inputs	Output	Inputs	Output
A B		A B	
0 0	0	1 1	1
0 1	1	1 0	0
1 0	1	0 1	0
1 1	1	0 0	0

truth table is simply an inversion of the positive truth table, as shown in Table 13-6.

Often in the course of digital electronics circuit design, you may encounter a complex looking truth table, that might appear to require a large number of gates. For example, consider the truth

Table 13-5. A Positive Logic NOR Gate Becomes a NAND Gate in Negative Logic.

Positive		Negative	
Inputs	Output	Inputs	Output
A B		A B	
0 0	1	1 1	0
0 1	0	1 0	1
1 0	0	0 1	1
1 1	0	0 0	1

table shown in Table 13-7. This would not be a very easy truth table to generate, or so it would seem. You might come up with something like the seven-gate circuit shown in Fig. 13-1. This will work, but it's awkward.

Table 13-6. The Positive and Negative Truth Tables for an Exclusive-OR Gate Are Shown Here.

Positive		Negative	
Inputs	Output	Inputs	Output
A B		A B	
0 0	0	1 1	1
0 1	1	1 0	0
1 0	1	0 1	0
1 1	0	0 0	1

Inputs				Output
A	B	C	D	
0	0	0	0	0
0	0	0	1	0
0	0	1	0	0
0	0	1	1	0
0	1	0	0	0
0	1	0	1	1
0	1	1	0	1
0	1	1	1	1
1	0	0	0	0
1	0	0	1	1
1	0	1	0	0
1	0	1	1	1
1	1	0	0	0
1	1	0	1	1
1	1	1	0	1
1	1	1	1	1

Table 13-7. Generating This Positive Logic Truth Table Could Appear to be Difficult (See Text).

Now, let's convert the same truth table to negative logic, as in Table 13-8. A simpler solution now becomes more obvious. This truth table can be generated with the four-gate circuit of Fig. 13-2. The circuits shown in these two figures are functionally identical. They generate the same truth table in either positive or negative logic.

Of course, it is possible to arrive at the simple solution of Fig. 13-2 directly from the positive logic truth table of Table 13-7. Using the negative logic form, simply makes the solution easier to find.

Table 13-8. The Negative Logic Truth Table for the Example Described in the Text Is Much Easier to Work With than the Positive Logic Version.

Inputs				Output
A	B	C	D	
0	0	0	0	0
0	0	0	1	0
0	0	1	0	0
0	0	1	1	1
0	1	0	0	0
0	1	0	1	1
0	1	1	0	0
0	1	1	1	1
1	0	0	0	0
1	0	0	1	0
1	0	1	0	0
1	0	1	1	1
1	1	0	0	1
1	1	0	1	1
1	1	1	0	1
1	1	1	1	1

Some designs will be easier with positive logic, and others will be simpler using negative logic. It is even okay to mix both positive and negative logic throughout the design of a single circuit, as long as you manage to keep them straight at all times. Negative logic is a handy tool for simplifying some designs. It is not essential, and some designers never bother with it. However, proper use of negative and positive logic will make life simpler for the digital circuit designer.

Chapter 14

Introduction to the Projects

We have dealt with a lot of information about a lot of digital electronics devices in the preceding chapters. Theoretical knowledge, however, can go just so far. The best type of education involves practical hands-on experience. In the remaining chapters of this book you will put what you have learned about using IC logic units to work in a number of actual projects that you can build and experiment with.

I very strongly recommend that you breadboard at least a few of the circuits discussed in the following pages. Actively working with digital components will give you a much stronger feel for the ways in which they function. Building these circuits will give you a firmer and more long-term education than just reading about them possibly can.

BREADBOARDING

Initially, at least, you should just breadboard the projects, rather than immediately soldering permanent versions of the circuits. Breadboarding is simply a process of temporarily hooking up a circuit for testing purposes. The term comes from the early days of radio, when experimenters would connect components with spring clips (or some other mechanical connector) on a slab of wood like a kitchen breadboard. The name stuck, even though today actual breadboards are no longer used for the purpose.

Most of today's experimenter's "breadboards" are solderless

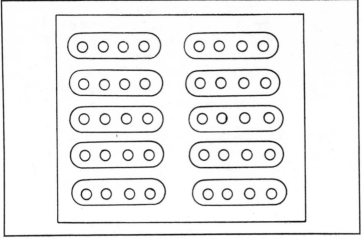

Fig. 14-1. The internal structure of a solderless breadboarding socket consists of rows of connected holes.

sockets, which allow a number of components to be easily, connected, but circuits can easily be changed without the nuisance of desoldering, and the components can be reused time after time.

The internal structure of a typical breadboarding solderless socket is shown in Fig. 14-1. Groups of holes, or contact points are shorted together in regular rows. Since two sets of rows are aligned with each other, this type of socket is perfect for use with DIP ICs. Each integrated circuit is mounted in the center of the socket to straddle two sets of contact rows, as illustrated in Fig. 14-2.

A number of breadboarding systems have been developed, and are available from a number of manufacturers. In addition to the

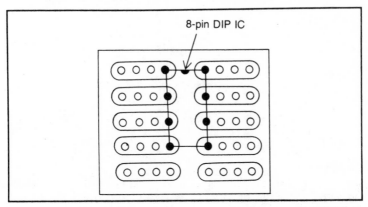

Fig. 14-2. Solderless breadboarding sockets are ideal for use with DIP ICs.

solderless socket, these systems include commonly used circuits such as power supplies, LED readouts, and signal generators. Large components, such as potentiometers, switches, and transformers, are also frequently included on breadboarding systems. A typical breadboarding system is shown in Fig. 14-3.

A large number of commercially sold breadboarding systems are available. Some experimenters prefer to build their own customized systems with the type of support circuitry their work most frequently calls for. In either case, a good breadboarding system is extremely valuable to anyone who works with electronics either professionally, or as a hobby. This is especially true for anyone doing any kind of design or circuit modification work. Different component values and circuit configurations can easily and quickly be changed until the desired results are achieved, without risking damage to any of the components.

BUILDING PERMANENT PROJECTS

If you choose to construct a permanent version of one or more of these projects, a solderless breadboarding socket would no longer be appropriate. These sockets do not hold components securely, and parts could fall out during use. Another problem is

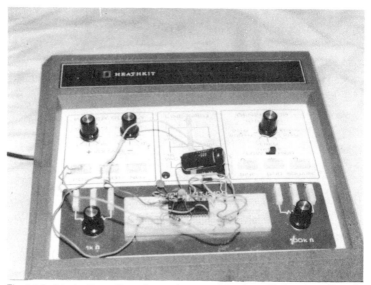

Fig. 14-3. A typical breadboarding system consists of a solderless breadboarding socket, and commonly used circuits such as power supplies and signal generators.

that moving the project about might cause the leads of two or more components to touch, creating a short.

Use of a solderless breadboarding socket often precludes ideal component layout and positioning. Noise pick-up, stray capacitances, and unwanted oscillations may become serious problems once you take a breadboarded circuit from the controlled conditions of the test bench. Finally, using a solderless breadboarding socket in your permanent projects adds an unnecessary and inefficient expense. It can also result in projects that are considerably more bulky than they need to be.

Solderless breadboarding sockets are designed for experimenting with and testing circuits on a temporary basis, and they are extremely well suited to such an application. However, they are a poor choice for projects intended for permanent, or long-term use. The two most common options open to the hobbyist constructing a permanent project are soldering and wire-wrapping.

Soldering

Until fairly recently, virtually all circuits were soldered together. That is, a mechanical connection is made between the component leads, and a special metal, principally tin and lead, is melted over the connection to hold it in place and to ensure the conduction of electricity through the connection. We will assume here that you are already familiar with the basics of soldering discrete electronic components. We will concentrate on the special requirements of soldering integrated circuits.

Some projects can be built on ordinary perfboard (perforated board). Of course, the perforations must be small and close to each other to accommodate the tightly spaced pins of DIP ICs. Connections can be made by physically twisting leads from the actual components directly together, or lengths of hook-up wire can be stretched from component to component.

The leads on DIP ICs are rather short, and so closely placed that twisting another lead or piece of wire around a single pin is often difficult. Also, there is the constant temptation to space components somewhat widely apart, and use moderately long lengths of hook-up wire between them. This certainly gives you more room to work in and makes the task of soldering easier, however, long leads and interconnecting wires can introduce all sorts of nasty problems, especially in the fairly high speed signals encountered in many digital electronics circuits. Printed circuit (or PC) boards are definitely to be preferred, especially for projects including more than two or three integrated circuit packages.

On a printed circuit board the connections between the components are made with traces of copper affixed to the board itself. Component leads are inserted through holes in the board and soldered directly to the copper pad, as shown in Fig. 14-4. Each soldered connection consists of only a single component lead and the copper pad affixed to the board. Leads can be cut off to very short lengths, and the already short leads of DIP ICs no longer pose a problem.

It is not difficult for the electronics experimenter to design and etch his own printed circuit boards. Starting with a solidly copper-clad board, the desired traces are drawn on with special ink, or strips of special tape. In some cases, the traces are applied by photographic techniques. In any case, the unwanted copper is etched, or eaten away by a special acid.

A number of beginner's kits and designers aids are available for designing and etching your own PC boards. I would especially recommend the tape/applique products with pre-made traces correctly spaced for various components. Since IC pins are placed very close to each other, manually drawing or painting a correctly proportioned set of PC pads is difficult at best. The tape traces can simply be peeled off their backing paper and affixed directly to the board.

Since integrated circuit pins are packed tightly next to each other (especially on DIP packages), great care must be exercised during soldering. It is very easy to create a solder bridge (a short circuit of excess solder) between adjacent pins.

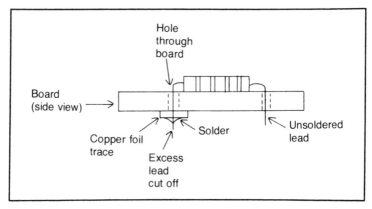

Fig. 14-4. Component leads are soldered directly to the foil trace on a printed circuit board, and the excess length of lead is cut off.

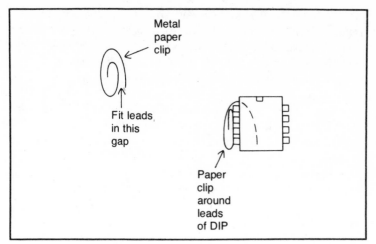

Fig. 14-5. A paper clip makes a convenient and inexpensive heatsink for soldering integrated circuits.

Another potential trouble area in soldering integrated circuits is the heat involved. ICs are semiconductor devices, and like all such components, they can be permanently damaged by excessive heat. The soldering iron tip should be applied just long enough to get the solder flowing.

A heatsink is a good idea. Special IC soldering heatsinks are commercially available, or you can just use a metal paper clip around the pins on each side of a DIP IC. This is illustrated in Fig. 14-5. Even with a heatsink, the soldering iron should never be in contact with the IC's leads for more than a few seconds at a time.

Wait a few seconds after soldering one lead before moving on to the next. This gives the chip time to cool down. It also allows the newly soldered joint time to set completely, which reduces the chances of it slopping over to merge with the next joint to be soldered, creating a solder bridge.

Another simple trick is to solder pins on opposite ends of the chip. The applied heat is not focused entirely on one spot this way. For example, the pins of a 14-pin DIP might be soldered in this order— 1 - 8 - 2 - 9 - 3 - 10 - 4 - 14 - 5 - 13 - 6 - 12 - 7 - 11.

Desoldering ICs can be extremely difficult because of the number of connections that have to be freed at once. If you have to desolder a good IC for some reason, you run a strong risk of destroying it with the heat of desoldering. Moreover, the task of removing the solder from all of those closely spaced pins is a major nuisance at best.

Fortunately, ICs rarely go bad, so you are not likely to need to replace them very often. However, it does happen from time to time. This makes the use of sockets for ICs highly desirable. Using sockets also removes the risk of damaging the IC while soldering, since only the socket is soldered to the board, and then the chip is inserted into the socket.

Some experimenters feel using sockets for all ICs is wasteful. Often the socket itself may be more expensive than the SSI (Small-Scale Integration) integrated circuit it is to replace. For many MSI (Medium-Scale Integration), and virtually all LSI (Large-Scale Integration) devices, this cost objection ceases to exist. It is inexpensive insurance to use a $1 socket to protect a $30 chip.

I feel, even for the less expensive ICs, sockets are still worthwhile insurance against the frustration you're likely to experience if you ever have to replace one. Of course, sockets are indispensable for applications where chips are likely to be changed from time to time. For example, in circuits using memories, we may want to use different ROMs or PROMs at various times (see Chapter 11).

If you choose to not use sockets, you must be especially careful when soldering the pins of CMOS ICs. The tip of the soldering iron must be solidly grounded to prevent potentially damaging static discharges (see Chapter 12 for more information). Battery operated or cordless rechargeable soldering irons may be used with reasonable safety on CMOS devices.

Wire-Wrapping

The other major method of permanently hooking up IC based circuits is wire-wrapping. For wire-wrapping, perfboard and special wire-wrap IC sockets may be used. The leads on these sockets are long, and have sharply squared off edges. A special wire-wrapping tool is used to wrap thin wire (typically about 30 gauge) around the leads, or posts (as they are sometimes called). The sharp edges bite into the wire, creating a firm electrical connection without soldering. If changes in wiring need to be made, the wire can usually be easily unwrapped. A wire-wrapped connection is illustrated in Fig. 14-6.

Wire-wrapping is subject to a number of disadvantages, which may or may not be of significance, depending on the individual application. First off, the sockets must be used, which adds to the project's total cost, of course. The special wire-wrapping sockets cost three or four times as much as regular IC sockets with an equivalent number of pins.

172

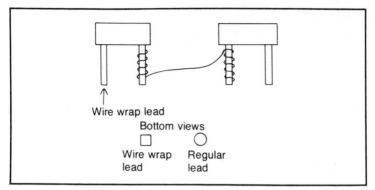

Fig. 14-6. Some hobbyists build their circuits using the wire-wrapping technique.

If the circuit to be constructed includes discrete components such as resistors and capacitors, their rounded leads can not be directly wire-wrapped. Some components may be inserted into wire-wrap sockets. Otherwise, they must be soldered. Soldering the thin wire-wrap wire is rather tricky. If just one or two discrete components are included in a circuit employing a large number of ICs, wire-wrapping may still be the way to go. However, in projects built around just one or two chips and a handful of discrete components, wire-wrapping is generally not very practical.

Wire-wrapping is also not a very good idea for projects that are intended to handle high frequencies. Wire-wrapped connections tend to filter out higher frequencies, and are somewhat subject to high frequency noise. The wire-wrapped connection is not as electrically sound as a well soldered connection. In most cases it will be more than adequate, but in critical applications there may be some problems.

Wire-wrapping a circuit is fairly quick, especially if an automatic wire stripping and wrapping tool is used (many are available from most distributors). Even when using a manual wrapping tool, the project can usually be assembled faster than soldering, and without the potential problems of soldering heat build-up. A wire-wrapped circuit can be easily and quickly rewired, if necessary. Also, since sockets are required, ICs can easily be changed whenever necessary.

Wire-wrap socket leads may be soldered, so some experimenters use them when constructing circuits point-to-point on regular perforated board. The longer lead of a wire-wrap socket is easier to work with than the short (printed circuit oriented) leads of regular IC sockets, and the ICs themselves.

PARTS SUBSTITUTION

In building one of these projects, you may want to substitute one component for another for any of a number of reasons. Of course, if you want to build a circuit and the parts called for aren't available, you'll want to look into substituting parts.

Sometimes you will want to make minor changes in a circuit. Often this will call for a few changes in component values. You may want to use a different logic family than the one called for in the project. In such an event, some parts substitutions are likely to be necessary, in addition to the changes in the ICs themselves.

In most of the projects, discrete component values are not too critical. Ordinary disc capacitors, electrolytic capacitors, and 5% or 10% ¼-watt carbon resistors may be used, unless noted otherwise. Often, changing the values slightly won't matter. For instance, a 0.47 μF and a 0.5 μF capacitor may be considered identical. Similarly, if a 12 k resistor is called for, a 10 k resistor, or a 15 k resistor may be substituted without changing the operation of the circuit in any significant way. When values are critical for some reason, this will be noted.

Component tolerances and power handling capacity can be increased in all cases. If a 10% resistor is called for, a 5%, or 1% unit may certainly be used. There will be no particular advantage in doing so, however. It will only increase the project cost. Of course, if you happen to have a supply of 5% resistors in your junk box, there is no reason for running out and buying 10% units, just because that is what is called for.

For most digital electronics work, ¼-watt resistors are used (½-watt or 1-watt resistors may be used with no problems). Higher-wattage components are larger and probably more expensive. In many of these circuits ⅛ watt resistors could be used. However, these components are fairly difficult to find at decent prices. The reduction in size will not be of any practical importance in hand-built versions of these projects. This may not be true of other projects you encounter from other sources, however.

It is not a good idea to use a lower wattage than what is called for in the parts list. In most digital electronics circuits, ⅛-watt resistors may be sufficient, but if the parts list calls for ¼-watt devices, you don't know if that's just what the designer happened to work with, or if the resistors actually need to dissipate ¼-watt. Do not derate component power handling unless you really know what you are doing.

Capacitors also have power ratings, given in the maximum

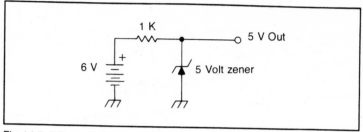

Fig. 14-7. A Zener diode can provide fair voltage regulation to drive TTL circuits from a six-volt battery.

amount of voltage the unit can withstand without breaking down. Once again, you can increase this rating. If a 25-volt capacitor is called for, a 50-volt, or even a 100-volt unit will work just as well, although at a potential increase in size and cost.

Do not use a lower voltage than the one called for in the parts list. If a circuit is powered by a 9-volt battery, a 10-volt capacitor may not be sufficient in some parts of the circuit, where the voltage might go higher than the supply voltage (by consuming current) for brief periods. For digital electronics circuits, either TTL or CMOS, 25-volts is a good minimum rating for capacitors.

If you exchange ICs, make sure the logic families match. For these projects either use all TTL or all CMOS. Interfacing families in these simple projects is likely to cause more trouble than it's worth. If you decide to change logic families for any reason, make sure you use the correct power supply. For some families, such as

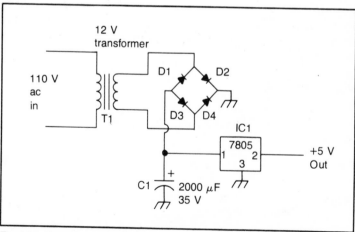

Fig. 14-8. This five-volt power supply circuit includes a voltage regulator for use with TTL circuits.

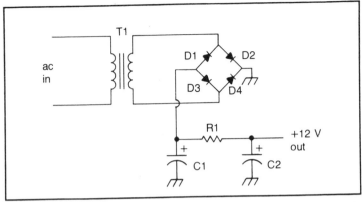

Fig. 14-9. CMOS circuits can be operated with this twelve-volt power supply circuit.

CMOS, all unused inputs to a chip should be grounded or connected to a logic 1 voltage to prevent instability.

Any time you substitute one IC for another, carefully compare the pinout sheets. If pinout diagrams are not available for both devices, do not make the substitution. If the function of just one pin is different, your project may not work properly, or at all. Moreover, you can damage the substituted chip, and possibly other ICs in the circuit too.

Generally, I'd recommend against changing IC type numbers, at least until you are very familiar with them. If you must change logic families, try to stay within the 74xxx numbering system. Like numbered devices (i.e., 7474, 74LS74, 74H74, and 74C74) have identical pinouts. Pinout diagrams for many of the most commonly used 74xxx and CD4xxx series devices are given in the Appendices for your convenience.

POWER SUPPLIES

Any electronic circuit requires some source of power. You could power CMOS digital projects with a 9-volt battery, and TTL circuits could operate on a 6-volt battery (four 1.5-volt cells),

Table 14-1. Here Is the Parts List for the 5-Volt TTL Power Supply Shown in Fig. 14-8.

IC1	7805 5 = volt regulator
D1 - D4	1N4001 diode
T1	110 volt to 12 volt (approximate) transformer
C1	2000 μF (or higher) electrolytic capacitor

D1 - D4	1N4001 diode
T1	110 volt to 12 volt (approximate) transformer
C1, C2	1000 μF (or higher) 35 volt electrolytic capacitor
R1	1 k ½ watt resistor

although a 5-volt zener should be used because of this family's
sensitivity to over-voltages. See Fig. 14-7. However, ac-to-dc
power supplies are often more convenient than batteries, especially
for breadboarding.

Figure 14-8 shows a 5-volt power supply that can be used to
drive any of the TTL projects presented in this book. The 7805
voltage regulator IC prevents problems with both over- and under-
voltages. The voltage supplied by this circuit is a very smooth 5
volts dc, even if the input voltage fluctuates.

The CMOS circuits may use the same power supply as the TTL
projects. However, CMOS devices work better at somewhat higher
voltages. The circuit in Fig. 14-9 is a 12-volt power supply that is
suitable for powering any small CMOS circuit. The parts lists for
the two power supplies are given in Tables 14-1 and 14-2.

Chapter 15

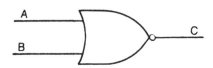

Simple Projects

In this chapter several fairly simple digital electronics projects will be presented and discussed. While fairly easy to build, these circuits can help you learn a great deal about how digital gates function.

LOGIC PROBE

Our first project will be a piece of test equipment that can be used to study and (if necessary) troubleshoot the rest of the projects presented in this book, and most other digital circuits too. The project is called a logic probe. Logic probes are simple but powerful tools for analyzing what goes on in a digital electronics circuit. Many commercially available logic probes with all sorts of extra features can be readily purchased. However, this inexpensive do-it-yourself logic probe will probably come in very handy for any digital electronics experimenter. What this device lacks in special features, it more than makes up for with low cost. The project should not cost you more than two or three dollars.

A logic probe is simply a device that allows you to determine the logic state (0 or 1) at any point in a digital circuit. A super-simple two component logic probe is shown in Fig. 15-1. The ground lead can be fitted with an alligator clip so that it can be connected to the same ground as that of the circuit being tested.

The probe is a short length of stiff solid wire, or a common test lead probe. If this probe is touched to a point in the circuit with a

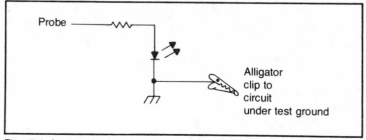

Fig. 15-1. A super simple logic probe can be made from a resistor and an LED.

logic 1 (high voltage) signal, the LED will light up. At all other times, the LED will remain dark.

The resistor is used for current limiting. If the LED is allowed to draw too much current it can be damaged. This resistor will generally have a value of less than (1000 ohms), with 330 ohms being a typical value. This super simple logic probe can be used without modification for any of the major logic families (CMOS, TTL or any of its variations, DTL, etc.). Since the circuit contains no active logic elements, and takes its power from the circuit being tested, it will be universal.

This circuit can easily be whipped together for under a buck. While functional, this super simple approach leaves a lot to be desired. For one thing, this simple circuit does not give a definite indication of a logic 0 state. If the LED does not light you may have a logic 0 signal, or the probe may not be making good contact with the pin being tested. There might even be a broken lead, or the LED itself could be damaged. With the circuit of Fig. 15-1 there is no way of telling.

Another potential problem area with this circuit is the possibility of excessively loading the IC output being tested. This can especially occur when the gate in question is already driving close to its maximum fan-out potential. Both of these problems can be side stepped by adding a pair of inverters and a second LED/ resistor combination, as illustrated in Fig. 15-2.

When a logic 0 is applied to the probe tip, the first inverter will change it to a logic 1, lighting LED A. The second inverter changes the signal back to a logic 0, so LED B remains dark. Conversely, when a logic 1 is fed into the probe tip, the output of the first inverter is a logic 0, so LED A stays dark, and LED B is lit up by the logic 1 signal appearing at the output of the second inverter.

This circuit can give a definite indication of either a logic 1 or a logic 0 signal. If neither LED lights up, the probe is not making proper contact. This circuit is much less ambiguous than the earlier version.

But that's not all this improved logic probe can tell you. This device can also indicate the presence of a pulsating signal that keeps reversing states. If the LEDs alternately blink on and off, a low-frequency pulse is being fed to the probe. If both LEDs appear to be continuously lit, a high-frequency pulse signal is indicated. Actually, in this case the LEDs are still alternately blinking on and off, but it is happening far too fast for the eye to catch it, so the LEDs look like they are both staying on at all times.

The circuit of Fig. 15-2 calls for just five components—two LEDs, two resistors (330 to 1000 ohms—the value is not critical), and a hex inverter IC. Two of the six inverter sections are used in this circuit. The other four inputs should be grounded to insure circuit stability. This is particularly important if a CMOS chip is being used.

Like the simpler version of the logic probe discussed earlier, this circuit steals its power from the circuit being tested. Alligator clips on the power supply leads can be attached to the power supply output of the circuit being tested.

A CMOS CD4009A hex inverter IC is the best choice, since it can be driven by either TTL or CMOS devices. By tapping into the tested circuit's power supply, the voltage levels will be automati-

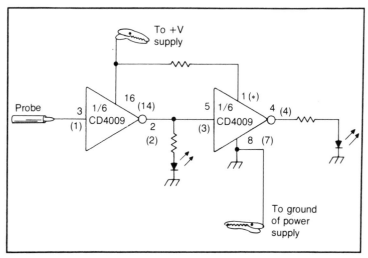

Fig. 15-2. An improved logic probe uses inverters to prevent circuit loading and to display both logic 1 and logic 0 signals.

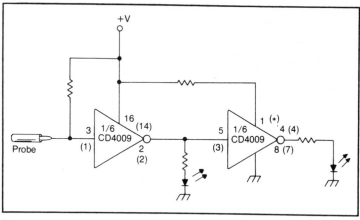

Fig. 15-3. If a pull-up resistor is included in your logic probe, a CMOS probe can be used on TTL circuits.

cally matched. In some cases, it may be desirable to add a pull-up resistor at the probe's input, as shown in Fig. 15-3. The value of this resistor is not critical, but it should probably be kept in the neighborhood of 1 k (1000 ohms).

If you intend to work with just TTL ICs, you can substitute a 7404 hex inverter (or the equivalent in the appropriate subfamily, like the low-power Schottky 74LS04). Use the pin numbers that are shown in parentheses in Fig. 15-2. The pin marked (∗) has no equivalent on the 7404, and should simply be ignored.

I strongly recommend that you build a permanent version of this project. It will be used throughout the rest of the projects for testing and circuit analysis.

DIGITAL CLOCK/OSCILLATORS

Many digital electronics circuits require some kind of clocking signal. Generally, this signal is obtained from a square-wave generator, or oscillator. Because of its function, a digital oscillator circuit is often referred to as a clock.

Many different clock circuits are available, depending on the exact requirements of the circuit to be clocked. Actually, this project will feature several different circuits. You may build any one or more of these devices.

A number of digital clocking circuits can be built around simple inverters. For example, the circuit in Fig. 15-4, for instance, is made up of three TTL inverters (half of a 7405 hex inverter IC), three resistors, and a capacitor. Resistor R may be anything from

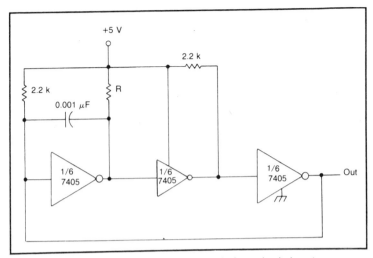

Fig. 15-4. A digital clocking circuit can be made from simple inverters.

about 1 k to about 4 k (1000 to 4000 ohms). The output frequency will be approximately 1 to 10 MHz (1,000,000 to 10,000,000 hertz), depending on the value of R.

Two independent clock circuits can be built using a single 7405 IC. Of course, this circuit can also be built around one of the TTL subfamilies, such as the 74L05, 74H05, 74LS05, or whatever. The 7405 hex inverter is used instead of the somewhat more common 7404 hex inverter, because the 7405 has what are called open collector outputs. This circuit may not function properly if a 7404 is used.

Figure 15-5 shows a clock circuit built around a pair of CMOS inverters (1/3 of a CD4049 hex inverter IC). The unused inverter inputs are grounded to insure stability. This circuit is extremely small. Only two components are required in addition to the IC itself—a capacitor and a resistor. These two discrete components determine the output clock frequency. The capacitor should be kept in the 0.01 μF to 10 μF range. A 500 k potentiometer can be used for R to make a variable frequency clock. If only a single frequency is needed, a fixed resistor can be used for R. The output frequency can be approximately determined with the following frequency:

$$F = \frac{1}{1.4\ RC}$$

where F is the frequency in hertz, R is the resistance in megohms (millions of ohms), and C is the capacitance in microfarads (μF). This equation is only an approximation, especially with high tolerance components. For example, if R = 1 k (1000 ohms, or 0.001 megohm), and C = 0.01 μF, the output frequency would be equal to 1/(1.4(0.001) (0.01)) = 1/0.000014 = approximately 71,500 Hz (71.5 kHz).

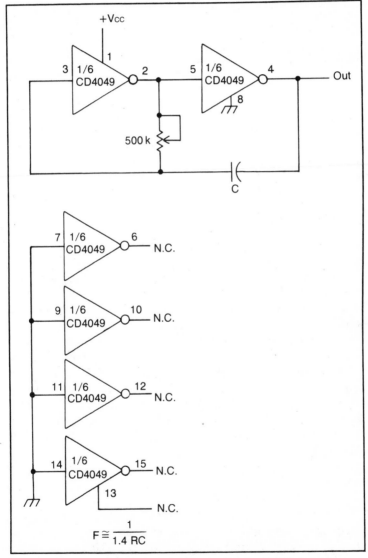

Fig. 15-5. Here is an inverter based clock circuit using CMOS devices.

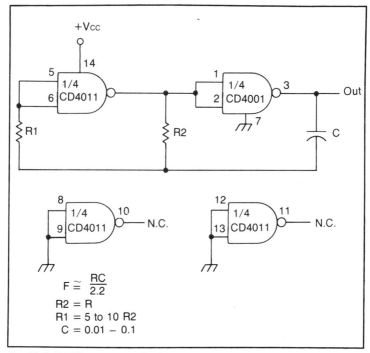

Fig. 15-6. NAND gates can also be used to make a digital oscillator.

To take a second example, we can set R to 500 k (0.5 megohms), and C to 10 μF. This gives us an output frequency of about $1/(1.4(0.5)(10)) = 1/7$, or just under 2 Hz. Plainly this circuit has a rather wide range of output frequencies.

A rather similar circuit is shown in Fig. 15-6. Here CMOS NAND gates (CD4011) are used in place of the inverters. A second resistor is also added to the circuit. Typically the capacitor will be somewhere between 0.01 and 0.1 μF. R2 will be between about 10 k (10,000 ohms) and 1 megohm (1,000,000 ohms), with R1 about 5 to 10 times the value of R2. R2 can be replaced with a potentiometer for a variable frequency clock. The approximate output frequency formula for this circuit is:

$$F = \frac{RC}{2.2}$$

Once again, this equation is just an approximation. Some typical component values for this circuit are given in Table 15-1.

An interesting variation on this circuit is shown in Fig. 15-7.

F	R1	R2	C
45 Hz	68 k	10 k	0.01 μF
450 Hz	680 k	100 k	0.01 μF
4.5 kHz	10 Meg	1 Meg	0.01 μF
450 Hz	68 k	10 k	0.1 μF
2 kHz	330 k	47 k	0.1 μF
4.5 kHz	680 k	100 k	0.1 μF
21.4 kHz	3.3 Meg	470 k	0.1 μF
45 kHz	10 Meg	1 Meg	0.1 μF

One of the NAND gate inputs can be externally controlled. A digital signal can turn the clock on or off. A logic 0 input inhibits the oscillator operation, while a logic 1 input enables it. Such a gated clock can come in handy in many advanced circuits.

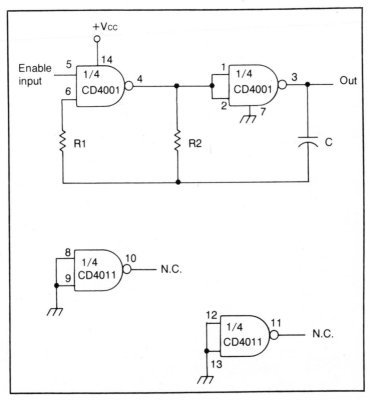

Fig. 15-7. The NAND gate clock can be gated by an external signal with a simple circuit modification.

NOR gates can also be used to create digital clocking circuits. The circuit illustrated in Fig. 15-8 is a combination digital/crystal oscillator. Of course the crystal is the primary factor in determining the output frequency. Varying the value of capacitor C (typically from 4 to 40 pF) allows some fine tuning of the output frequency for precision applications.

The component values indicated in the schematic are selected for peak performance around 1 MHz (1,000,000 Hz). For other output frequencies, you might want to experiment with the component values to get the cleanest possible output signal.

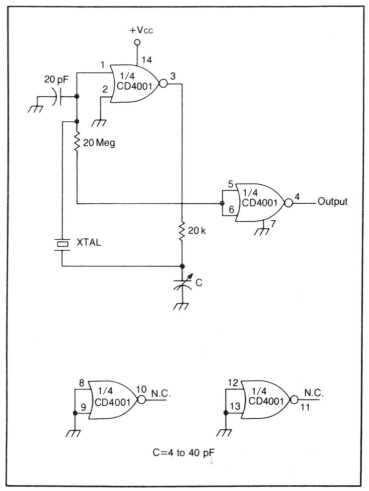

Fig. 15-8. Some digital gates use crystals for a precise output frequency.

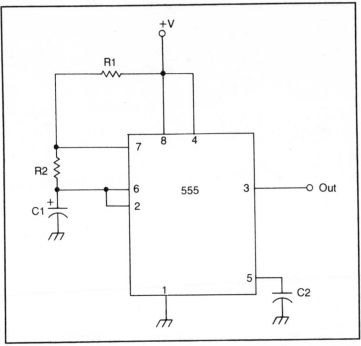

Fig. 15-9. Even though it is not a true digital device, the 555 timer is often used to clock digital circuits.

Many digital circuits use linear timer ICs such as the 555, or XR-2420 to provide clocking signals. These devices are easy to work with, inexpensive, and suitable for direct driving of most of the logic families. Figure 15-9 shows a 555 based clock, and Fig.

Table 15-2. The 555 Timer Clocking Circuit of Fig. 15-9 Can Use Many Different Resistor and Capacitor Combinations.

C1	R1	R2	F
0.01 μF	1 k	1 k	48 kHz
0.01 μF	1 k	10 k	6.9 kHz
0.01 μF	10 k	100 k	0.7 kHz
0.1 μF	1 k	1 k	4.8 kHz
0.1 μF	1 k	10 k	0.7 kHz
0.1 μF	4.7k	10 k	0.6 kHz
0.001 μF	1 k	1 k	480 kHz
0.001 μF	1 k	10 k	69 kHz
0.001 μF	2.2 k	15 k	45 kHz
0.001 μF	10 k	18 k	31 kHz
0.001 μF	33 k	100 k	6.2 kHz
1 μF	1 k	4.7 k	0.14 kHz

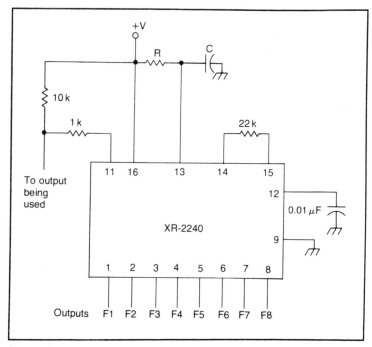

Fig. 15-10. Here is a clock circuit using the XR2240 programmable timer IC.

15-10 shows a multiple output clock built around the XR-2240 programmable timer. Also, see Tables 15-2 and 15-3.

The clock oscillator circuit of Fig. 15-9 requires only two resistors and two capacitors in addition to the 555 IC itself. The time for the logic 1 output and the time for the logic 0 output for each output cycle are determined separately. Both resistors and capacitor C1 determine the logic 1 output time, according to the following formula:

Table 15-3. The XR2240 Programmable Timer Can Cover an Extremely Wide Range of Output Frequencies.

R	C	F1	F2	F3	F4	F5	F6	F7	F8
1 k	0.01 μF	100 k	50 k	25 k	12.5 k	6.25 k	3125	1562	781
10 k	0.01 μF	10 k	5 k	2.5 k	1.25 k	625	312	156	78
1 k	0.05 μF	20 k	10 k	5 k	2.5 k	1.25 k	625	312	156
2.2 k	0.05 μF	9.1 k	4.5 k	2.3 k	1.1 k	568	284	142	71
10 k	0.05 μF	2 k	1 k	500	250	125	62	31	16
1 k	0.1 μF	10 k	5 k	2.5 k	1.25 k	625	312	156	78
4.7 k	0.1 μF	2.1 k	1.1 k	532	266	133	66	33	17
10 k	0.1 μF	1 k	500	250	125	62	31	16	8

(All frequencies are in Hz. unless marked with a k for kHz).

188

$$T_1 = 0.693 \times (R1 + R2) \times C1$$

The logic 0 output time is dependent only on resistor R2 and capacitor C1:

$$T_0 = 0.693 \times R2 \times C1$$

The value of the second capacitor (C2) is not involved in the timing calculations. This capacitor will usually have a value of 0.01 μF.

The total time of each entire cycle is found simply by adding the logic 1 and logic 0 output times:

$$T = T_1 + T_0$$

This can be converted to frequency by taking the reciprocal of the time period:

$$F = \frac{1}{T} = \frac{1}{0.693(R1 + R2)C1 + 0.693\ R2\ C1} = \frac{1.44}{(R1 + R2)C1}$$

Some typical values are shown in Table 15-2.

The XR2240 programmable timer is a more complex device than the 555, but, paradoxically, it's generally easier to work with. For one thing, the equation for determining the output frequency is much more straightforward. In fact, it is about as simple as such a formula could possibly get:

$$F = \frac{1}{RC}$$

The unique thing about this chip is that in addition to the basic timer circuitry, this device includes a built-in timer counter, or frequency divider. Pins 1 through 8 are all outputs that may be used independently or in combination to generate a wide range of frequencies with a single resistor/capacitor combination.

The output from pin 1 is the basic frequency, determined by the above formula. We will call this frequency F_1. The other pins put out binary factors of this frequency:

Pin #2 — $F_2 = F_{1/2}$
Pin #3 — $F_3 = F_{1/4}$

$$Pin \#4 - F_4 = F_{1/8}$$
$$Pin \#5 - F_5 = F_{1/16}$$
$$Pin \#6 - F_6 = F_{1/32}$$
$$Pin \#7 - F_7 = F_{1/64}$$
$$Pin \#8 - F_8 = F_{1/128}$$

For example, if $C = 0.01 \mu F$ and $R = 10$ k, the output frequency for each pin will be as follows:

$$F_1 = 10,000 \text{ Hz}$$
$$F_2 = 5000 \text{ Hz}$$
$$F_3 = 2500 \text{ Hz}$$
$$F_4 = 1250 \text{ Hz}$$
$$F_5 = 625 \text{ Hz}$$
$$F_6 = 312.5 \text{ Hz}$$
$$F_7 = 156.25 \text{ Hz}$$
$$F_8 = 78.125 \text{ Hz}$$

That is an extremely wide range for a single device. Some additional examples are given in Table 15-3. Any output that is used should be connected to the positive power supply through a 10 k

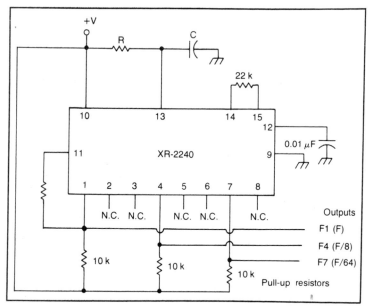

Fig. 15-11. The XR2240 timer offers multiple output frequencies, if pull-up resistors are used.

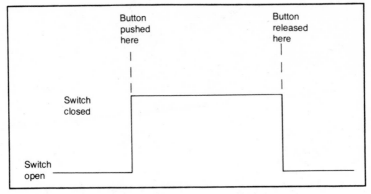

Fig. 15-12. An ideal switch would cleanly open and shut.

pull-up resistor, as illustrated in Fig. 15-11. Any or all of the circuits presented in this section would make suitable clocks for digital circuits.

BOUNCE-FREE SWITCH

Mechanical switches are not perfect. Rather than neatly making contact, as shown in Fig. 15-12 when they are closed, they tend to bounce open and shut several times very rapidly before settling into position, as shown in Fig. 15-13. For most analog applications

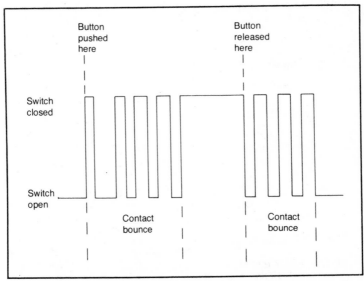

Fig. 15-13. Practical switches tend to suffer from contact bounce.

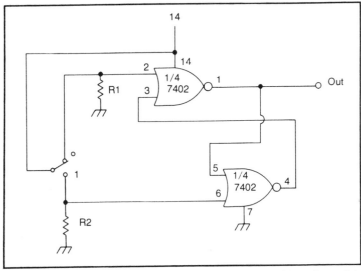

Fig. 15-14. This circuit can be used to eliminate the problem of bouncing switch contacts.

this scarcely matters. The bouncing and settling in period are far too rapid to even be noticeable. Digital circuits, however, are designed to recognize very brief pulses, so bouncing switch contacts could be a problem in many circuits. For instance, let's say we have a digital counter set up to keep track of the number of times the button is pushed. Each bounce of the contacts would look like a separate switch closure to the counter. Instead of counting one push of the switch as 1 (which is what we want), it will count a random number of bouncing pulses.

We expect to get 0 - 1 - 2 - 3 - 4 - 5 - 6 - 7 - 8 - 9 - 0 - 1 - 2 - 3 - 4 - 5 - 6 - 7 - 8 - 9 - 0 - 1 - 2 - 3 - 4 . . . But what we end up getting, is something like this: 0 - 6 - 1 - 4 - 0 - 3 - 8 - 2 - 5 - 9 - 2 - 7 - 1 - 6 - 0 - 4 - 0 - 3 - 8 - 2 . . . Obviously this could be a problem that might render a digital circuit completely useless. What we need is some way to get the digital circuitry to ignore the unwanted bouncing contact pulses.

Table 15-4. The Switch Debouncing Circuit of
Fig. 15-14 Can Be Constructed from These Components.

IC1	½ (2 sections)	7402 quad NOR gate (or 74L02, 74H02, 74S02, 74LS02, 74C02, etc.)
R1,R2	100 k resistor	
S1	SPDT switch	

A good way to do this is to have the first closure of the contacts trigger a monostable multivibrator (see Chapter 6) with an output time that is longer than the bouncing (settle-in time) of the switch. A circuit for accomplishing this task is shown in Fig. 15-14 and the parts list is given in Table 15-4.

As shown as the switch contacts first make contact, the one-shot is triggered. Further switch openings and closures will be ignored until the multivibrator completes its timing period. This period does not have to be long (by human standards). A fraction of a second will be sufficient.

The way the bounce-free switch works is illustrated in Fig. 15-15. Bounce-free switches will not be necessary for every manually operated switch in every digital circuit, but when the need arises, this simple circuit can make the difference between a functional, high-technology piece of equipment, and a useless piece of junk.

LED FLASHER

LED flashers have been popular for simple projects ever since electronics experimenters got into digital techniques. In fact, the flasher project predates the digital revolution. Analog flashers that featured blinking flashlight bulbs or neon lights were once common in the pages of the electronics hobbyist magazines. The LED flasher is a modernization of this old stand-by of hobbyists.

The popularity of these flasher projects may seem a bit odd, since these circuits don't really do anything practical. They simply

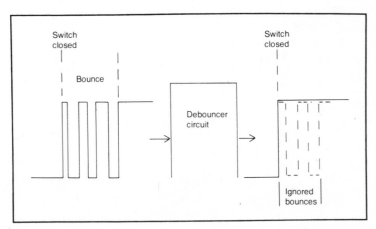

Fig. 15-15. A monostable timer ignores the multiple opens and shuts of bouncing switch contacts.

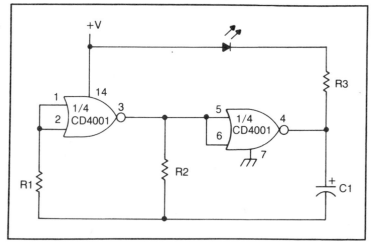

Fig. 15-16. This simple circuit flashes a LED on and off at a regular rate.

turn a light source (such as a LED) on and off in a regular pattern. But a LED flasher project has quite a bit going for it. It is a simple and inexpensive project that rewards the hobbyist by doing something obvious after an hour or so of work.

Practical uses for LED flashers can be found. They make fine, eye-catching displays. They can also be incorporated into electronic toys (for instance, a pair of flashing LEDs could be a robot's blinking eyes), or they can be put to more practical work as alarm indicators. The flashing LED is highly visible and attention grabbing, making it an ideal indicating device.

A simple LED flasher circuit built around two CMOS NOR gates is shown in Fig. 15-16. The parts list is given in Table 15-5. Try using different values for R2 and C1 for different blinking rates. Making either (or both) of these components smaller will cause the LED to flash faster. If the flashing rate is made too fast (more than 6 to 10 times a second), the LED will appear to be constantly lit,

**Table 15-5. Just a Handful of Parts Are Needed
to Build The Simple LED Flasher Circuit of Fig. 15-16.**

IC1	CD4001 quad NOR gate (two sections left unused)
D1	LED
R1	1 megohm ¼ watt resistor
R2	180 k (see text) ¼ watt resistor
R3	820 ohm ¼ watt resistor
C1	10 μF 35 volt electrolytic capacitor (see text)

because the human eye can not react fast enough to catch the individual blinks.

This circuit should look fairly familiar. It's a variation on the digital oscillators discussed earlier in this chapter. The main difference is that the frequency determining components are selected for a very slow rate of oscillation. The components shown in the parts list will give a blinking rate of about 1 Hz (or one flash per second).

Resistor R3 is used to limit the current being drawn through the LED, to protect it from damage. Changing the value of this resistor will change the brightness of the LED when it is on.

By adding another NOR gate stage, a fourth resistor and a second LED, as illustrated in Fig. 15-17, we can convert the circuit to an alternate LED flasher. The LEDs will blink in turn. When LED 1 is lit, LED is dark, and vice versa. The LEDs will never both be simultanously lit (although they may appear to both be constantly lit at high flashing rates), and the only time they will both be dark is when there is no power applied to the circuit. This is accomplished by using the third gate section as an inverter to reduce the original output signal (which is fed to LED 1) and feed the inverter signal to the second LED.

This modification should have very little impact on the project cost, since the CD4001 IC contains four NOR gates in a single

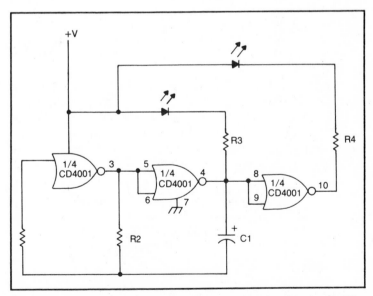

Fig. 15-17. Adding another gate section allows the LED flasher circuit of Fig. 15-16 to alternately flash two LEDs.

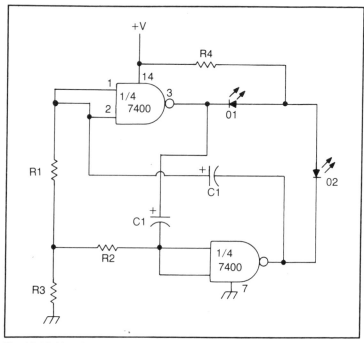

Fig. 15-18. Here is another circuit for alternately flashing a pair of LEDs.

package. The only new components are R4 and LED 2. If you don't have these components in your junk box, it might cost you about 35¢ to buy them. R4 should have the same value as R3, otherwise the brightness of the two LEDs will look unbalanced.

Another alternate LED flasher circuit is shown in Fig. 15-18. This time 7400 NAND gates are used. This circuit will work with TTL (7400), any of its subfamilies (74L00, 74S00, 74LS00, 74H00), or CMOS (74C00). The gates in this project are arranged as a self-trigging flip-flop, or astable multivibrator (see Chapter 6). The flash rate is determined by R1, C1, R2, and C2. If R1 = R2, and C1

Table 15-6. The Parts List for the Second
Alternate LED Flasher Circuit Shown in Fig. 15-18.

IC1	7400 quad NAND gate (or 74L00, 74LS00, 74S00, 74H00, 74C00 — see text)
C1, C2	100 μF 25 volt electrolytic capacitor (see text)
R1, R2	3.9 k ¼ watt resistor (see text)
R3	1 k ¼ watt resistor
R4	470 ohm ¼ watt resistor
D1, D2	LED

196

= C2 then each LED will be on for an equal amount of time. The on times for the two LEDs can be unbalanced by making these component values unequal (Table 15-6).

Resistor R4 protects both LEDs from excessive current. Since the two LEDs are never on simultaneously, the single resistor may be used. Separate current-limiting resistors (as in Fig. 15-17) could also be used with no change in the circuit's operation. The two techniques were shown in the two circuits for purposes of illustration.

In the next chapter we will go to work on somewhat more complex projects built around digital counter circuits.

Chapter 16

Counter Projects

One of the functions that digital electronics circuits are particularly well suited for is counting. After all, digital circuits, by definition, deal with data in digital, or numerical form. What more natural use can we put numbers to than counting?

Counting may seem like a rather simple and limited application for digital electronics, but nothing could be further from the truth. While the basic principles are not complicated, there are a great many different basic approaches to digital electronics counting, and far more practical applications.

In this chapter we will explore a number of varied digital counter circuits and discuss a few of their applications. Several projects will be presented for you to build and experiment with.

BINARY COUNTERS

Since digital electronics components work with the binary numbering system, the simplest counter circuits are binary counters. For demonstration purposes, the binary counter projects presented in this section will use LEDs as output displays. A lighted LED represents a binary 1, and a dark (unlighted) LED will stand for a binary 0.

Be sure to use a current-dropping resistor in series with any LEDs in any of these projects. Without a series resistor, an LED can draw an excessive amount of current, possibly damaging itself and other components in the circuit. A 10-cent resistor is certainly cheap protection, and it will hold the current through the LED down to an acceptable level. Series resistors will be shown in the schematic diagrams for the various projects in this chapter, but values will not be specified. The reason for this is that the exact value of the resistor is not particularly critical. Making the resistance smaller will cause the LED to burn brighter, or increasing the resistance will result in a dimmer glow. For best results, the resistor should have a value between 220 ohms and 2.2 k (2200 ohms). Resistors with values smaller or larger than this range should not be used.

Of course, if one of these counter circuits is used as part of a larger system, the outputs may drive other digital gates, and a visible display will not be needed. Obviously, this would mean that the LED and its current-dropping resistor can be eliminated and the output of the counter circuit can be fed directly to the input of the next circuit stage.

Single-Digit Binary Counter

Unquestionably the simplest digital counter circuit would be a single-digit binary counter. That is, one that counts from 0 to 1, and that starts over at 0 again. The counting sequence would be 0 - 1 - 0 - 1 - 0 - 1 . . .

A simple D-type flip-flop can be used to accomplish this. Half of a 7474 (or 74C74, 74LS74, or whatever) dual-D flip-flop integrated circuit is used in the circuit shown in Fig. 16-1. The $\overline{Q}$ (NOT Q)

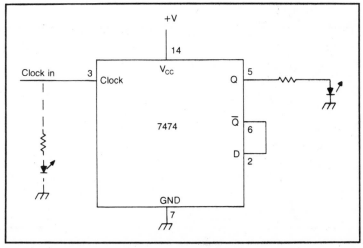

Fig. 16-1. A flip-flop can be used as a single digit binary counter.

complementary output is fed back into the D (data) input. The main output (Q) will reverse states after each complete input cycle (see Fig. 16-2).

If you want a clearer idea of how this circuit works, you can add an input display as shown in dotted lines in Fig. 16-1. The input can be any of the digital clock circuits described in the last chapter. To make the blinking of the LEDs visible, the clock should have a very low frequency—1 to 3 hertz, or less. If a high frequency clock is used, the LEDs will appear to be continuously lit, even though they are actually blinking on and off at a high rate. If you build this demonstrator circuit with LED 1 displaying the input state, and LED 2 displaying the output state, you should see something like this pattern:

LED 1	LED 2	Output Count
off	off	0
on	off	0
off	on	1
on	on	1
off	off	0
on	off	0
off	on	1
on	on	1
off	off	0
on	off	0
off	on	1
on	on	1

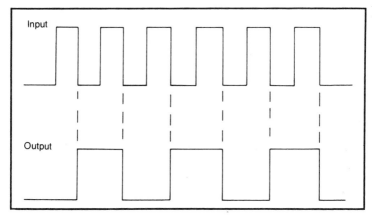

Fig. 16-2. The input and output signals for a single-digit binary counter demonstrate how it can function as a frequency divider.

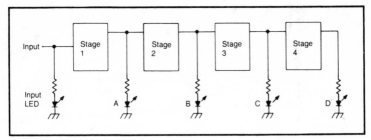

Fig. 16-3. Cascading flip-flops makes a more useful multi-digit binary counter.

and so forth. The circuit counts complete input cycles.

Plainly a device that can only count from 0 to 1 is of rather limited value. It is shown here to demonstrate the basic principles involved.

The principle application of this type of circuit is as a frequency divider. The output frequency will be exactly one half of the input frequency. For instance, if the clock is running at 1200 hertz (1.2 kHz), then the output frequency will be 600 hertz.

Multidigit Binary Counters

Just as the simple gates discussed in the first few chapters in this book could be greatly increased in their capabilities and applications, binary counters can also be combined to create more useful multidigit binary counters.

The basic principle of multidigit binary counters is illustrated in the block diagram of Fig. 16-3. The output of each flip-flop stage show the same pattern of the single-digit binary counter discussed above. The end result is the counting sequence shown in Table 16-1. A four-stage binary counter can count from 0000 (decimal 0) to 1111 (decimal 15), before resetting and starting over. Such a circuit can count up to 16 input pulses.

Figure 16-4 shows a more detailed example. This is a two stage counter using a pair of J-K flip-flops for a count sequence of:

00
01
10
11
00
01
10
11
00

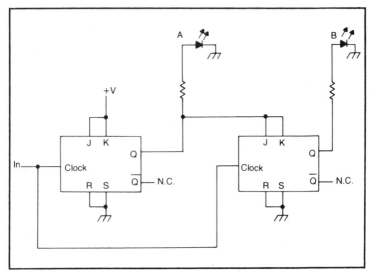

Fig. 16-4. This is the circuit for a practical two-stage binary counter using JK flip-flops.

The count sequence repeats after every four input pulses.

The output of stage 1 feeds the J and K inputs of stage 2. As you should recall, when the J and K inputs are held low there will be no change in output state when a clock pulse is received. When the J and K inputs are at logic 1, the output will reverse states (change from a 0 to a 1, or from a 1 to a 0) after every complete clock pulse received.

Since the J and K inputs of the first stage are permanently tied to the positive power supply line (if you are using TTL devices, a resistor between 390 ohms and 22000 ohms should be added between the inputs and +V), this stage will behave like a single-digit binary counter, reversing its output stage once after every complete input cycle.

If the output of stage 1(Q1) is at 0, the J and K inputs of stage 2 will be effectively grounded (held at logic 0), so this stage will ignore any clock pulses while in this condition—its output will remain constant until the output of stage 1 (Q1) goes to 1, forcing the output of stage 2 to reverse its state on the next clock pulse. This gives us the four-step counting pattern described above.

Figure 16-5 shows a practical circuit of this type using a single IC—the CMOS CD4027 dual J-K flip-flop. Notice that except for the IC itself, the two display LEDs and their current dropping resistors, no additional components are required. A similar circuit can be built around the TTL 7473 dual J-K flip-flop, or any of its derivatives (74L73, 74H73, 74C73, 74LS73, etc.), although the pin numbers will have to be changed.

By feeding back the $\overline{Q}$ (NOT Q) output of stage 2 to the J input of stage 1, we can force the counter to reset after every three clock pulses, changing the count sequence to:

$$00$$
$$01$$
$$10$$
$$00$$
$$01$$
$$10$$
$$00$$
$$01$$

and so forth.

Notice that both LEDs will never be lit (the output count will never be 11). Either one or both of the LEDs will be dark (equal to 0) at all times. By adding a NOR gate across the two outputs, as shown in Fig. 16-6, we can change the circuit from a 3-step binary counter to a 3-step sequential counter. LED A will be lit when the output

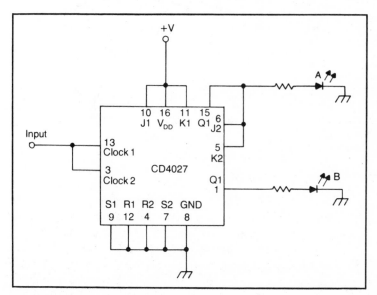

Fig. 16-5. Here is a practical version of the two-stage binary counter shown in Fig. 16-4.

203

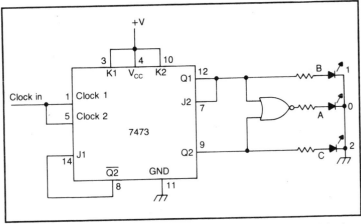

Fig. 16-6. Feedback can be used for odd counts, such as this modulo-three counter.

count is 00, LED B will be lit when the output count is 01, and an output count of 10 will light LED C. Only one LED will be lit at any given instant. The counting sequence now looks like this;

A	B	C
1	0	0
0	1	0
0	0	1
1	0	0
0	1	0
0	0	1
1	0	0
0	1	0
0	0	1
1	0	0

and so forth.

Either with or without the NOR gate (since this circuit has three possible output counts) it is called a modulo-three counter. The unmodified two-stage circuit of Fig. 16-5 is a modulo-four counter, since it cycles through a sequence of four counting steps.

The modulo can be increased by adding more stages. A four-stage counter like the one illustrated in Fig. 16-3 is a modulo-sixteen counter with sixteen counting steps (refer back to Table 16-1).

Input LED	LED A	LED B	LED C	LED D	Binary Count	Decimal Equivalent
off	off	off	off	off	0000	0
on	off	off	off	off	0000	0
off	on	off	off	off	0001	1
on	on	off	off	off	0001	1
off	off	on	off	off	0010	2
on	off	on	off	off	0010	2
off	on	on	off	off	0011	3
on	on	on	off	off	0011	3
off	off	off	on	off	0100	4
on	off	off	on	off	0100	4
off	on	off	on	off	0101	5
on	on	off	on	off	0101	5
off	off	on	on	off	0110	6
on	off	on	on	off	0110	6
off	on	on	on	off	0111	7
on	on	on	on	off	0111	7
off	off	off	off	on	1000	8
on	off	off	off	on	1000	8
off	on	off	off	on	1001	9
on	on	off	off	on	1001	9
off	off	on	off	on	1010	10
on	off	on	off	on	1010	10
off	on	on	off	on	1011	11
on	cn	on	off	on	1011	11
off	off	off	on	on	1100	12
on	off	off	on	on	1100	12
off	on	off	on	on	1101	13
on	on	off	on	on	1101	13
off	off	on	on	on	1110	14
on	off	on	on	on	1110	14
off	on	on	on	on	1111	15
on	on	on	on	on	1111	15
off	off	off	off	off	0000	0
on	off	off	off	off	0000	0
off	on	off	off	off	0001	1
on	on	off	off	off	0001	1
off	off	on	off	off	0010	2
on	off	on	off	off	0010	2

(the cycle repeats from 0000 to 1111)

A practical modulo-sixteen counter circuit using two CD4013 dual D-type flip-flop ICs is shown in Fig. 16-7. Compare this circuit with those using the J-K type flip-flop stages. The two techniques function in essentially the same way, since the D input is like a tied together set of J and K inputs. By adding more stages, and/or using

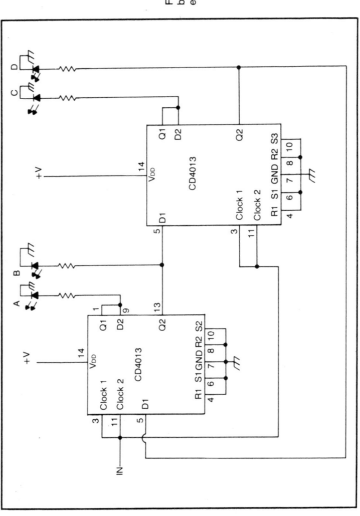

Fig. 16-7. Four D-type flip-flops can be cascaded to create a modulo-eight binary counter.

feedback to reset after a specific count, a binary counter with any desired modulo higher than two can be created. (A modulo-1 counter would be worthless since it would have only one possible output state which would never change. A modulo-0 counter would be even more worthless—it would have no possible outputs!)

SEQUENTIAL COUNTERS

A sequential counter has multiple outputs, of which only one is active at any one time. Some designs have all outputs at logic 0, except for the active output, which is at logic 1. Other circuits reverse this, with the active output being pulled low (logic 0), while all of the other outputs are high (logic 1).

We have already worked with one sequential counter, the modulo-three counter shown in Fig. 16-6. In this section we will work on some projects involving sequential counters with larger modulos.

The CD4017 decade counter/divider IC is an extremely versatile device that can be used to create sequential counters of almost any modulo. As the pinout diagram of Fig. 16-8 shows, the CD4017 has ten sequential outputs, numbered from 0 to 10. Only the activated output will be at logic 1. The other nine will be held at

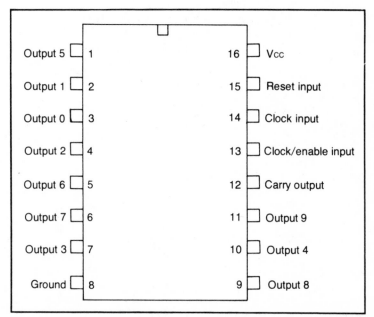

Fig. 16-8. The CD4017 is a sequential decade counter in IC form.

Table 16-2. The Modulo-Seven
Sequential Counter of Fig. 16-9 Activates Outputs 0 Through Six in Turn.

Input	Outputs									
Pulse #	1	2	3	4	5	6	7	8	9	0
1	1	0	0	0	0	0	0	0	0	0
2	0	1	0	0	0	0	0	0	0	0
3	0	0	1	0	0	0	0	0	0	0
4	0	0	0	1	0	0	0	0	0	0
5	0	0	0	0	1	0	0	0	0	0
6	0	0	0	0	0	1	0	0	0	0
7	0	0	0	0	0	0	0	0	0	1
8	1	0	0	0	0	0	0	0	0	0
9	0	1	0	0	0	0	0	0	0	0
10	0	0	1	0	0	0	0	0	0	0
11	0	0	0	1	0	0	0	0	0	0
12	0	0	0	0	1	0	0	0	0	0
13	0	0	0	0	0	1	0	0	0	0
14	0	0	0	0	0	0	0	0	0	1
15	1	0	0	0	0	0	0	0	0	0

(This sequence is continuously
repeated).

logic 0. This chip is a component modulo-ten sequential counter. Since it has ten counting steps, it is called a decade counter.

The CD4017 can easily be wired as a sequential counter with a modulo less than ten simply by connecting pin #15 (reset) to the pin representing the highest desired count. For example, the circuit

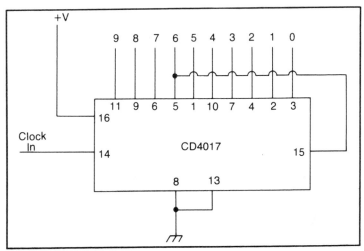

Fig. 16-9. The CD4017 can be wired for any modulo from two to ten.

shown in Fig. 16-9 will count up to 6, then recycle back to 0 and start over. The counting sequence for this circuit is given in Table 16-2. Notice that outputs 7, 8, and 9 are never activated (raised to logic 1). The count jumps from 6 to 0, then 1-2-3, etc. Since the counting sequence ranges from 0 to 6, this device is a modulo-7 sequential counter.

Obviously other modulos less than ten can be set up in the same way, simply by tying the appropriate output pin (desired modulo less one) to pin #15.

Figure 16-10 illustrates a simple modification of the same basic circuit. By connecting pin #13 (clock enable) to the desired maximum count and grounding pin #15 (reset), the circuit will count up to that maximum count, and then stop. For instance, if the CD4017 is wired as shown in Fig. 16-10, the counting sequence will be: 1 - 2 - 3 - 4 - 5 - 6 - 7.

Once a count of 7 has been reached the counter will stop, ignoring any additional clock pulses. The circuit can be reset back to the initial count by temporarily placing a logic 1 at pin #15 (connecting this pin to +V). When pin #15 is grounded, the counter will step through its counting sequence once and only once, until it is reset again.

Input clock pulse are ignored if pin #13 is at logic 1. When the

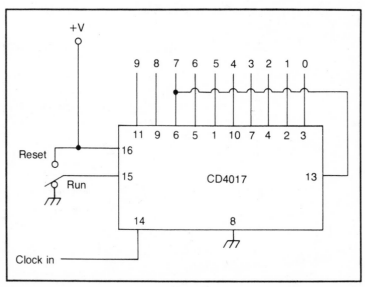

Fig. 16-10. A minor change in the circuit of Fig. 16-9 allows the CD4017 to count through only a single cycle.

desired maximum count is reached, it jumps to logic 1, disabling the chip from counting additional input pulses. In the continuously recycling version of the CD4017 circuit (Fig. 16-9), pin #13 is permanently grounded, so that the clock input pulses are recognized and counted at all times. Pin #15 resets the counter to its initial minimum count when a logic 1 is applied. In Fig. 16-9, when the desired maximum count is reached, it forces pin #15 high, causing the count to jump to zero. In Fig. 16-10 the logic signal fed

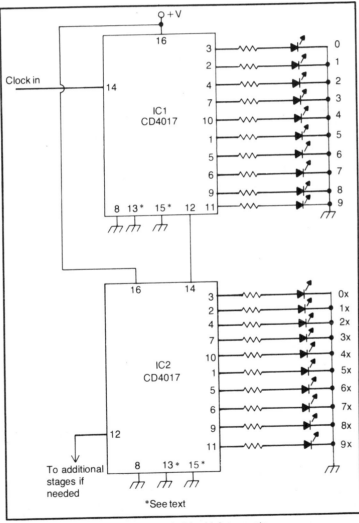

Fig. 16-11. CD4017's can be cascaded for higher counts.

Input	Outputs - IC1										Outputs - IC2									
Pulse #	1	2	3	4	5	6	7	8	9	0	0x	1x	2x	3x	4x	5x	6x	7x	8x	9x
1	1	0	0	0	0	0	0	0	0	0	1	0	0	0	0	0	0	0	0	0
2	0	1	0	0	0	0	0	0	0	0	1	0	0	0	0	0	0	0	0	0
3	0	0	1	0	0	0	0	0	0	0	1	0	0	0	0	0	0	0	0	0
4	0	0	0	1	0	0	0	0	0	0	1	0	0	0	0	0	0	0	0	0
5	0	0	0	0	1	0	0	0	0	0	1	0	0	0	0	0	0	0	0	0
6	0	0	0	0	0	1	0	0	0	0	1	0	0	0	0	0	0	0	0	0
7	0	0	0	0	0	0	1	0	0	0	1	0	0	0	0	0	0	0	0	0
8	0	0	0	0	0	0	0	1	0	0	1	0	0	0	0	0	0	0	0	0
9	0	0	0	0	0	0	0	0	1	0	1	0	0	0	0	0	0	0	0	0
10	0	0	0	0	0	0	0	0	0	1	0	1	0	0	0	0	0	0	0	0
11	1	0	0	0	0	0	0	0	0	0	0	1	0	0	0	0	0	0	0	0
12	0	1	0	0	0	0	0	0	0	0	0	1	0	0	0	0	0	0	0	0
13	0	0	1	0	0	0	0	0	0	0	0	1	0	0	0	0	0	0	0	0
14	0	0	0	1	0	0	0	0	0	0	0	1	0	0	0	0	0	0	0	0
					*	*	*	*	*	*	*	*	*	*	*	*				
87	0	0	0	0	0	0	1	0	0	0	0	0	0	0	0	0	0	0	1	0
88	0	0	0	0	0	0	0	1	0	0	0	0	0	0	0	0	0	0	1	0
89	0	0	0	0	0	0	0	0	1	0	0	0	0	0	0	0	0	0	1	0
90	0	0	0	0	0	0	0	0	0	1	0	0	0	0	0	0	0	0	0	1
91	1	0	0	0	0	0	0	0	0	0	0	0	0	0	0	0	0	0	0	1
92	0	1	0	0	0	0	0	0	0	0	0	0	0	0	0	0	0	0	0	1
93	0	0	1	0	0	0	0	0	0	0	0	0	0	0	0	0	0	0	0	1
94	0	0	0	1	0	0	0	0	0	0	0	0	0	0	0	0	0	0	0	1
95	0	0	0	0	1	0	0	0	0	0	0	0	0	0	0	0	0	0	0	1
96	0	0	0	0	0	1	0	0	0	0	0	0	0	0	0	0	0	0	0	1
97	0	0	0	0	0	0	1	0	0	0	0	0	0	0	0	0	0	0	0	1
98	0	0	0	0	0	0	0	1	0	0	0	0	0	0	0	0	0	0	0	1
99	0	0	0	0	0	0	0	0	1	0	0	0	0	0	0	0	0	0	0	1
100	0	0	0	0	0	0	0	0	0	1	1	0	0	0	0	0	0	0	0	0
101	1	0	0	0	0	0	0	0	0	0	1	0	0	0	0	0	0	0	0	0
102	0	1	0	0	0	0	0	0	0	0	1	0	0	0	0	0	0	0	0	0
103	0	0	1	0	0	0	0	0	0	0	1	0	0	0	0	0	0	0	0	0

(and so forth)

to this pin is determined by a manual switch. An external digital gate could also be used to feed the reset input of the CD4017.

This chip also has a carry output pin (pin #12), which allows multiple counters to be cascaded for higher counts. For instance, in Fig. 16-11, IC1 counts the ones (0-9), and IC2 counts the tens (0x to 9x). The carry output (pin #12) of IC1 is fed into IC2 as the clock input (pin #14). Everytime counter IC1 passes from 9 to 0, an output pulse is fed into the second counter, causing it to increment 1. Notice that for this circuit, two LEDs are lit at all times—one to represent the ones column and the other to represent the tens. For example, if pin #4 of IC1 is at logic 1 (output 2) and pin #1 of IC2 is also at logic 1 (output 5), the total count will be $2 \times 1 + 5 \times 10$, or 52. The counting sequence for this two-digit sequential decade counter circuit is illustrated in Table 16-3. To save space the entire sequence is not listed.

A third CD4017 (hundreds) can be added by connecting pin #12 (carry out) of IC2 to pin #14 (clock input) of the new IC3. The rest of IC3 would be wired in the usual fashion. This would extend the count range from 000 to 999. Any number of counters can be cascaded in this way.

By connecting the desired maximum count to pin #15 (for

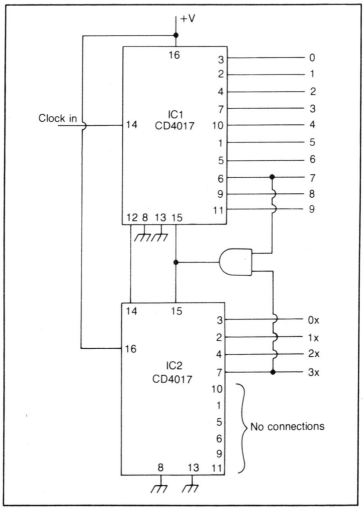

Fig. 16-12. Here is a circuit for a modulo-38 sequential counter using a pair of CD4017 ICs.

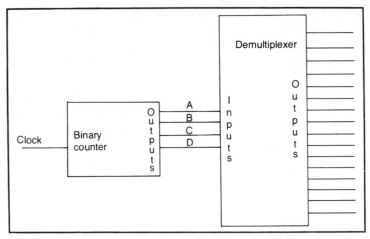

Fig. 16-13. Combining a binary counter with a demultiplexer is another way of creating a sequential counter.

automatic recyling) or pin #13 (for one cycle/manual reset) of all of the counters, any modulo up to the maximum count can be created. AND gates are used to set the reset value. Figure 16-12, for example, is a continuously recyling modulo-38 counter. It counts up to 37, then recycles back to 00. The counting sequence is as follows:

00 - 01 - 02 - 03 - 04 - 05 - 06 - 07 - 08 - 09 - 10 - 11 - 12 - 13 - 14 - 15 - 16 - 17 - 18 - 19 - 20 - 21 - 22 - 23 - 24 - 25 - 26 - 27 - 28 - 29 - 30 - 31 - 32 - 33 - 34 - 35 - 36 - 37 - 00 - 01 - 02 - 03 - 04 - 05 - 06 - 07 - 08 - 09 - 10 - and so on.

Clearly the CD4017 is a powerful device for a multitude of counter applications.

Sequential counters can also be created by combining binary counters with demultiplexers, as illustrated in the block diagram of Fig. 16-13. This is more practical than using separate gates (as in Fig. 16-6) for most modulos. The appropriate sequential output can be used to reset the binary counter for nonstandard modulos. The 74154 4-to-16 demultiplexer IC is fine for this kind of application. Of course, the CMOS version (74C154) or one of the TTL subfamilies (74L154, 74H154, 74LS154 . . .) could be readily substituted.

This 24-pin chip (see Fig. 16-14) works with the reversed output system mentioned earlier. That is, the active output is logic 0, while the other outputs are held at logic 1. The 74154 accepts a four-bit binary number (from 0000 to 1111) and activates the appropriate sequential output (0 to 15). If a four-stage binary counter is used as the input to this device, the output will sequentially step through its sixteen outputs:

0 - 1 - 2 - 3 - 4 - 5 - 6 - 7 - 8 - 9 - 10 - 11 - 12 - 13 - 14 - 15 - 0 - 1 - 2 - 3 - and so on.

A typical circuit is shown in Fig. 16-15. A four-stage flip-flop binary counter drives the demultiplexer for a 16-step sequential output series. If the binary counter is set up for a modulo less than 16, the output sequence will also be limited. For instance, if a modulo-10 binary counter is used, the 10, 11, 12, 13, 14, and 15 outputs of the demultiplexer will never be activated. Using LED outputs, a moving dot effect can be achieved. The light will appear to move down the line of LEDs. Try arranging the LEDs in a circle for a continuous motion effect.

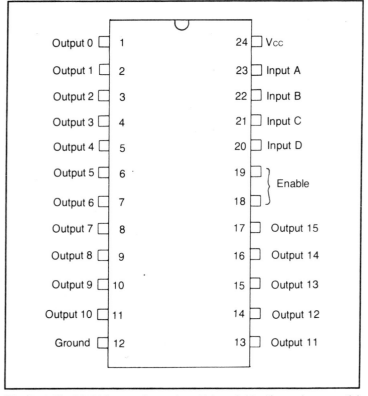

Fig. 16-14. The 74154 four-to-sixteen demultiplexer is ideal for use in sequential counters.

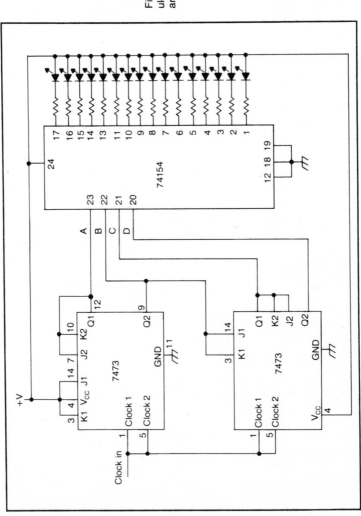

Fig. 16-15. Here is a practical modulo-sixteen sequential counter built around the 74154 demultiplexer.

While not actually a counter circuit, an interesting variation on this circuit is shown in Fig. 16-16. Four independent clocks are fed into the four inputs of the 74154 demultiplexer. Since the oscillators run at different rates, the LED outputs will be lit in a more or less random pattern.

A suggested parts list for this random flasher project is shown in Table 16-4. The components marked with asterisks (*) set the frequencies of the various oscillators, and may be freely changed. For best results, these component values should be relatively large (so the frequencies will be low enough to produce visible flashes), and different for each oscillator to keep the output pattern random.

UP/DOWN COUNTERS

Most counter circuits count upwards from 0 to the maximum count. That is, a four-bit binary counter counts:

0000
0001
0010
0011
0100
0101
0110
0111
1000
1001
1010
1011
1100
1101
1110
1111

Sometimes, however, it would be more useful to have a circuit that can count down from its maximum value to zero, like this:

1111
1110
1101
1100
1011
1010

```
1001
1000
0111
0110
0101
0100
0011
0010
0001
0000
```

In other words, the counting sequence is generated backwards.

For flip-flop based binary counters, the reverse counting effect can be readily achieved by using the $\overline{Q}$ (NOT Q) outputs instead of the regular Q outputs. This is illustrated in Fig. 16-17. Compare this circuit with the one shown in Fig. 16-6. Each output is now the exact opposite of what it would be in a standard counter, producing the count-down sequence.

The 74193 is a dedicated counter chip that can count in either direction (up or down) depending on the logic signals fed to pins 4 and 5. The pinout diagram for this device is shown in Fig. 16-18.

A logic 1 at the clear pin (#14) forces the outputs to 0000, independent of the previous count. The counter can also be preset to any value set at inputs A-D (pins #15, 1, 10, and 9) when the load input (pin #11) is set to logic 0. If pin #14 is grounded, and pin #11 is connected to the V+ line (through a resistor, if necessary), the device will behave like a standard 4-bit binary counter, with the

Table 16-4. Here Is the Recommended Parts List for the Random Flasher Circuit of Fig. 16-16.

IC1 - IC4	555 timer
IC5	74154 demultiplexer
R1, R2	470 k *
R3, R5	220 k *
R4, R7	100 k *
R6, R8	82 k *
R9 - R24	330 ohm (typical)
C1, C5	4.7 μF 25 volt electrolytic capacitor *
C3, C7	10 μF 25 volt electrolytic capacitor *
C2, C4, C6, C8	0.01 μF disc capacitor

(* - - - see text)

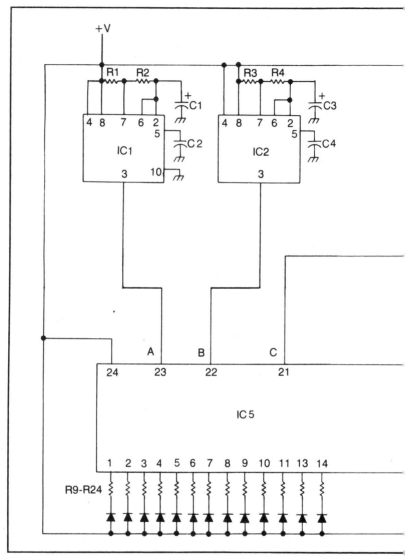

Fig. 16-16. Using four independent clocks to drive a 74154 four-to-sixteen demultiplexer creates a fascinating pseudo-random flasher.

direction of the count determined by the logic signals at pins 4 and 5. These pins are used as clock inputs. The count is incremented up one each time a logic 1 is fed to pin #5. Similarly, the output count is decremented down 1 for each high pulse fed to pin #4. Obviously this is an extremely versatile device. It is available in the various

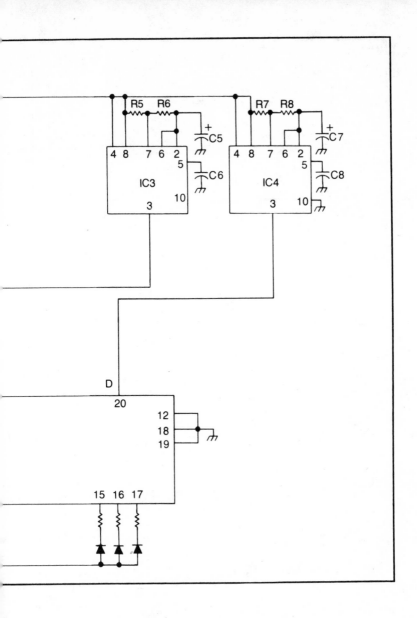

TTL subfamilies (74LS193 et al) too.

A novel pseudo-random output pattern can be achieved by feeding two different clock frequencies to pins #4 and 5, as illustrated in Fig. 16-19. The most dramatic effects can be achieved if the two input frequencies are widely spaced and do not bear a

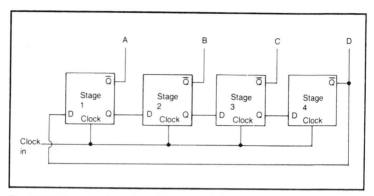

Fig. 16-17. Here is a four-stage binary count-down counter.

harmonic (multiple) relationship to each other.

The 74193 can also be used as an input source for a demulti-plexer such as the 74154. An interesting circuit along these lines is shown in Fig. 16-20. This circuit will count up, then reverse direction and count back down. That is, the output sequence will run like this: 0 - 1 - 2 - 3 - 4 - 5 - 6 - 7 - 8 - 9 - 10 - 11 - 12 - 13 - 14 - 15 - 14 - 13 - 12 - 11 - 10 - 9 - 8 - 7 - 6 - 5 - 4 - 3 - 2 - 1 - 0 - 1 - 2 - 3 - 4 - 5 - 6 - 7 - 8 - 9 - 10 - 11 - 12 - 13 - 14 - 15 - 14 - 13 - 12 - and so forth.

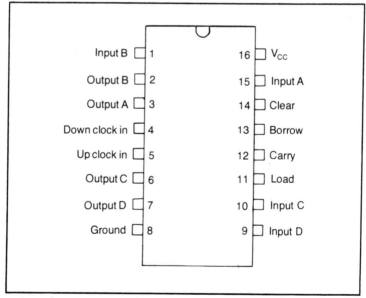

Fig. 16-18. The 74193 is a single chip binary counter that can count in either direction.

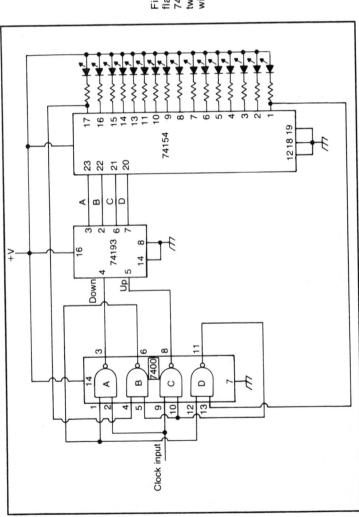

Fig. 16-19. A pseudo-random flasher can be built around the 74193 up/down counter by using two independent clock sources with widely separated frequencies.

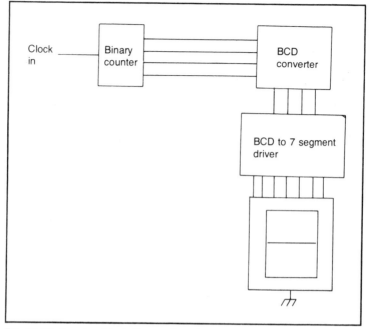

Fig. 16-20. Here is a block diagram for a basic decimal counter.

DECIMAL-OUTPUT COUNTERS

We human beings are generally used to dealing with numbers using the ten digits of the decimal system (0, 1, 2, 3, 4, 5, 6, 7, 8, and 9) not dots that are either on or off with their position determining their value. Hooking up a binary counter with a BCD (binary-coded-decimal) decoder and a seven-segment display driver will accomplish these ends readily. Other methods can also be employed.

For instance, consider the single-digit decimal number counter circuits shown in Figs. 16-21 and 16-22. The CD4518 is a dual BCD counter in a single IC package, eliminating the need for a separate binary counter and BCD converter. The CD4511 converts the BCD values into the format required for lighting the appropriate display segments (see Table 16-5). A TTL version using the 7490 BCD counter (sometimes called a decade counter) and the 7448 BCD to 7-segment decoder/driver. Either of these circuits can be expanded to more than one digit. Figure 16-23 illustrates a two-digit cascaded 7490/7448 circuit.

You should not feed the outputs of a regular binary counter into a BCD to 7-segment decoder like the CD4511 or 7448. These

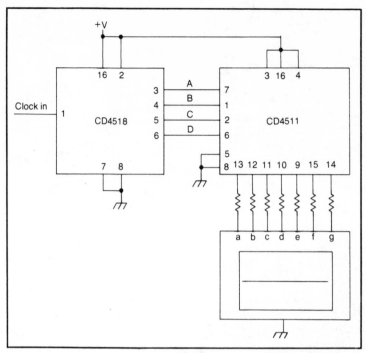

Fig. 16-21. This decimal counter circuit is built around the CD4518 and CD4511 ICs.

Table 16-5. A BCD to Seven-Segment Decoder Allows the Appropriate Decimal Digit to be Displayed.

Inputs				Outputs							Decimal
A	B	C	D	a	b	c	d	e	f	g	
0	0	0	0	x	x	x	x	x	x	-	0
0	0	0	1	-	x	x	-	-	-	-	1
0	0	1	0	x	x	-	x	x	-	x	2
0	0	1	1	x	x	x	x	-	-	x	3
0	1	0	0	-	x	x	-	-	x	x	4
0	1	0	1	x	-	x	x	-	x	x	5
0	1	1	0	x	-	x	x	x	x	x	6
0	1	1	1	x	x	x	-	-	-	-	7
1	0	0	0	x	x	x	x	x	x	x	8
1	0	0	1	x	x	x	-	-	x	x	9
1	0	1	0	Invalid							
1	0	1	1	Invalid							
1	1	0	0	Invalid							
1	1	0	1	Invalid							
1	1	1	0	Invalid							
1	1	1	1	Invalid							

223

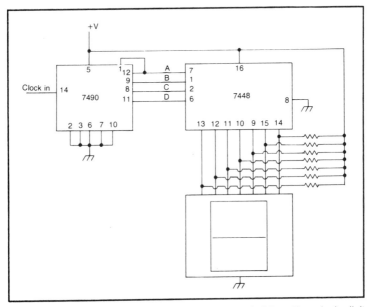

Fig. 16-22. The 7490 and 7448 ICs can be used to build another single-digit decimal counter.

devices are designed to accept only BCD data. Binary counters generate six combinations that are undefined in the BCD format:

$$1010$$
$$1011$$
$$1100$$
$$1101$$
$$1110$$
$$1111$$

If these binary combinations are fed into a BCD to 7-segment decoder, the output will be meaningless. In fact, some ridiculous nondigit is likely to be displayed.

INPUT SIGNALS

In all of the counter circuits described so far in this chapter, we have assumed that the input signal being counted is the regular signal generated by a clock or oscillator. While this is sometimes useful, the applications are limited. Fortunately digital counting circuits aren't too fussy about where the input signal comes from. The input pulses do not have to arrive at a regular rate. Many

224

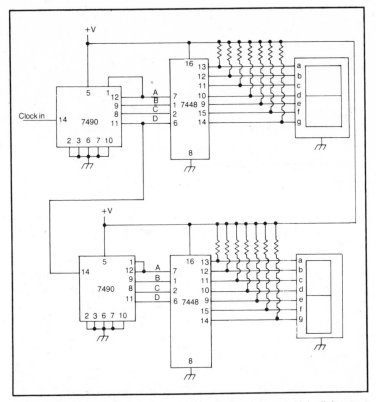

Fig. 16-23. Decimal counters can be cascaded to produce a multiple digit output and higher counting range.

applications call for manual input through a mechanical switch. Figure 16-24 shows how a SPDT switch can select between a logic 0 or a logic 1 input.

Simple SPST switches can also be used, if a resistor is added. For instance, in Fig. 16-25, if the switch is open, the input will be at logic 0. Closing the switch puts a logic 1 into the circuit. The

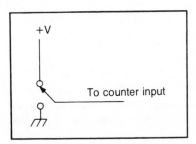

Fig. 16-24. A SPDT switch can be used to manually select the logic level being fed to an input.

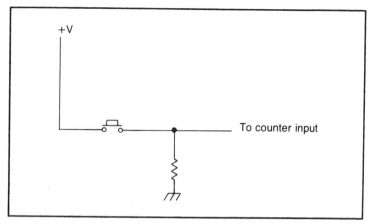

Fig. 16-25. Adding a resistor allows a SPST switch to be used to manually enter logic levels. This switch is a normally low version.

resistor prevents the V+ from being shorted directly to ground. The value of this isolation resistor is not critical. I generally use a 1 k (1000 ohm) resistor, but this is mainly due to habit.

The same technique can be applied for the opposite arrangement. That is, when the switch is open, the input is logic 1, but when the switch is closed, the signal goes to logic 0. This is illustrated in Fig. 16-26.

Of course any time you use mechanical switches, switch bounce may be a problem. The switch debouncing circuit described in Chapter 15 would be useful here to prevent accidental multiple counts for a single switch closure.

Mercury switches could be used in some applications. A mercury switch is a small glass tube with two electrical contacts inserted into one end. A small globule of mercury is sealed into the

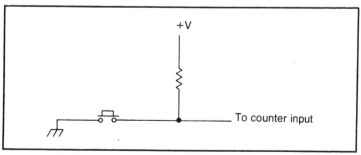

Fig. 16-26. Changing the position of the resistor converts the switch to a normally high unit.

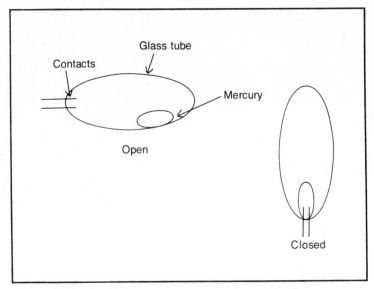

Fig. 16-27. A mercury switch senses its own position.

tube. When the tube is tilted so the mercury slides away from the contacts, the switch is open. Tilting the switch so that the mercury covers both of the electrical contacts the switch is closed. Thus, a mercury switch is a position sensor (Fig. 16-27).

Reed switches might also come in handy for certain counter applications. A normally-open reed switch is closed when a magnetic field is brought near it. A normally-closed reed switch is just the opposite—it opens in the presence of a magnetic field. A reed

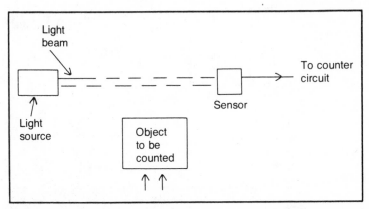

Fig. 16-28. A light source and a photosensitive device can act as the input for an object counter.

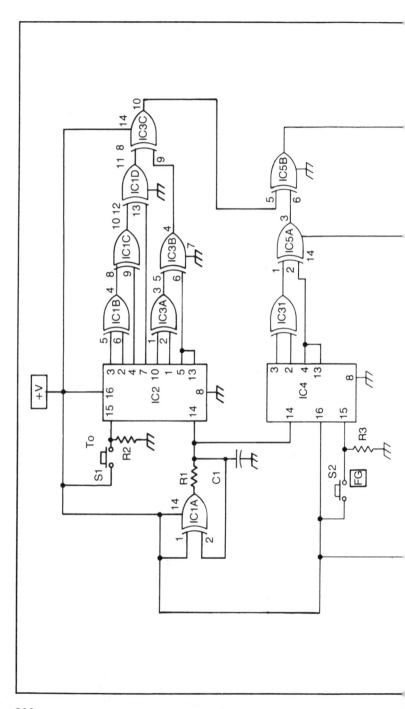

228

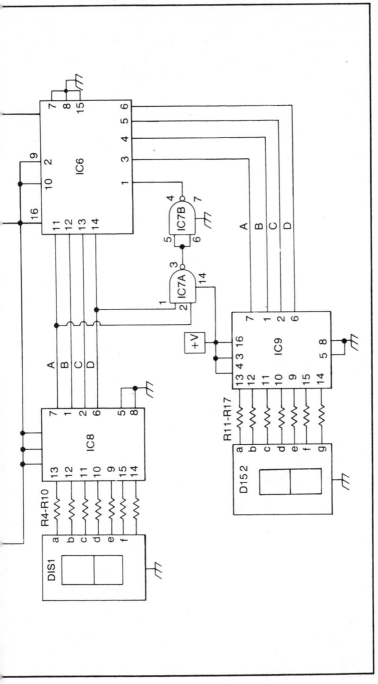

Fig. 16-29. This circuit can be used for automatic scoring for a football team.

Table 16-6. Here Is the Parts List for the Football Scorekeeper Project of Fig. 16-29.

IC1, IC3, IC5	CD4017 quad Exclusive-OR gate
IC2, IC4	CD4017 decade counter
IC6	CD4518 dual BCD counter
IC7	CD4011 quad NAND gate
IC8, IC9	CD4511 BCD to 7 segment decoder
DIS1, DIS2	common cathode 7 segment LED display
R1	4.7 k resistor
R2, R3	1 k resistor
R4 - R17	330 ohm resistor
C1	0.005 μF capacitor
S1, S2	SPST Normally Open push-button switch

switch can be used to count how many times a magnet is moved past a specific test point. Since a conductor with electric current passing through it produces a magnetic field, a reed switch can also be used to monitor the action of a separate circuit without any direct electrical connection.

Photosensitive devices also make good inputs for counters in object counting applications. See Fig. 16-28. When an object on a conveyor belt, or a person walking by temporarily blocks the sensor from the light source, the counter is incremented (or decremented, depending on the circuit). Of course logic signals from other digital gates and circuits can also be counted with an appropriate counter circuit.

The applications are virtually limitless. With a little bit of imagination, an input arrangement can be found to allow your circuit to count almost anything.

SCOREKEEPER PROJECT

A natural practical application for a digital counter would be a scorekeeping device for games and sporting events. When a player scores, the appropriate button is pushed to increment the counter.

However, many games do not score with even increments of one. For example, in football, a touchdown scores six, and a field goal counts as three. The operator could manually press the switch six or three times, but this is a nuisance at best, and there is always the risk of losing count of how many times the switch has been pushed so far. Clearly, this pretty much defeats the purpose of an automatic scorekeeping circuit.

A good solution would be to use multiple counters within the single circuit. Here is where a single cycle/manual reset counter, like the CD4017 circuit shown in Fig. 16-10 comes in handy. When the button is pushed, the single cycle counter is reset and run to feed the appropriate number of pulses into the main counter.

A complete football scoring unit for one team is shown in Fig. 16-29. Switch S1 is closed for a touchdown (6 points) and switch S2 is closed for a field goal (3 points). IC1 and IC2 are simply debouncing circuits to keep the input signals clean. Incidentally, do not assume that switch contact bounce will do the job for you—the number of bounces is irregular and unpredictable. The parts list for the scorekeeper project is shown in Table 16-6. Of course the circuit can easily be adapted for virtually any game by setting up the CD4017 single cycle counters for the appropriate unit count. The same idea can be applied to count-down counters.

Chapter 17

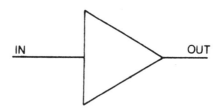

More Counter Projects

In the last chapter we have explored basic counter circuits in some depth. In this chapter we will continue working with digital counter circuits in more deluxe projects.

RANDOM NUMBER GENERATORS

Occasionally you may have need for a digital circuit that generates a random (or pseudo-random) number. This will be most useful for games. For instance, later in this chapter we will construct a set of electronic dice. Other applications may also call for random number generation. These include statistical experiments, random lights displays, and random music makers, among others.

Probably the easiest way to generate a random number in a digital circuit is to use a high-speed clock that is fed to a counter while a button is held closed, but stops when the switch is released. If the clock speed is high enough, all of the LEDs will appear to be constantly lit, and it will be impossible for the operator to predict what the final count will be when the button is released.

Figure 17-1 is a block diagram that demonstrates the simplicity of this approach. The schematic for a practical circuit using three CMOS ICs is shown in Fig. 17-2. The parts list is given in Table 17-1.

A 555 timer is used to generate the high-frequency clock signal. The frequency, as determined by R1, R2, and C1 is about 48 kHz (48,000 Hz) for the values given in the parts list. Four D-type

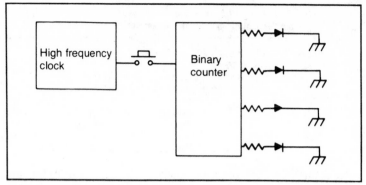

Fig. 17-1. This block diagram demonstrates how random binary numbers may be generated.

flip-flops (two CD4013 dual D flip-flop ICs) are connected as a four-stage binary counter. When pushbutton S1 is closed, the clock signal is fed into the binary counter. When the button is released, the counter will stop and display its last count.

This is one digital circuit where a switch debouncing circuit is unnecessary. If the switch bounces and a few more clock pulses get through to the counter, what difference does it make? We *want* a randomized output anyway.

The same approach can be used to display randomized decimal digits on a seven-segment LED display. A typical circuit for this is shown in Fig. 17-3. The parts list is given in Table 17-2. It works in the same way as the binary version of Fig. 17-2, except the flip-flop binary counter is replaced with a CD4518 BCD counter and a CD4511 BCD to 7-segment LED display. While the switch is closed, all seven segments of the display will appear to be constantly lit, so the readout will appear to be a solid 8. However, once the switch is released, the last count will be displayed. As with most other counters, additional stages can be cascaded to allow for higher

Table 17-1. Here Is a List of the Components
Required for the Binary Random Number Generator Project Shown in Fig. 17-2.

IC1	555 timer
IC2, IC3	CD4013 dual D flip-flop
D1 - D4	LED
R1, R2	1 k resistor
R3 - R6	330 ohm resistor
C1, C2	0.01 μF capacitor
S1	SPST (normally open pushbutton switch)

**Table 17-2. These Parts Are Used to Build
the Decimal Random Number Generator of Fig. 17-3.**

IC1	555 timer
IC2	CD4518 BCD counter
IC3	CD4511 BCD to 7 segment decoder
DIS1	common cathode 7 segment LED display
R1, R2	1 k resistor
R3 - R9	330 ohm resistor
C1, C2	0.01 μF capacitor
S1	SPST (normally open pushbutton switch)

counts. Cascading two of the decimal circuits will allow for random counts from 00 to 99.

There are two possible ways to cascade decimal number generators. The approach shown in Fig. 17-4 is essentially the same as for a two-digit decimal counter. When counter 1 passes from 9 to 0, a clock pulse is fed into the second counter (tens). Alternatively,

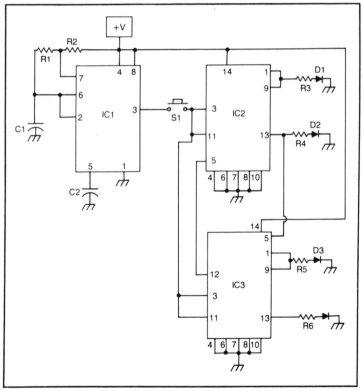

Fig. 17-2. Random binary numbers may be generated using this practical circuit.

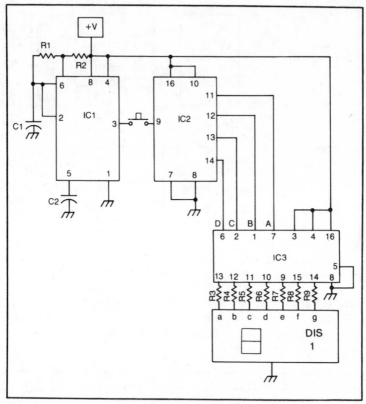

Fig. 17-3. This circuit will generate random decimal numbers.

separate clocks could be used to drive the two stages, making for a more random effect. This is illustrated in Fig. 17-5.

In many games, play is determined by the roll of dice. Why not a set of electronic dice? This is just a variation on the random number counters described above, except the count for each "die" ranges from 1 to 6, and is displayed in the traditional dice pattern.

The circuitry for a single die is shown in Fig. 17-6. A CD4017 decade counter chip is connected as a modulo-six cycling counter, with the gates of IC3 and IC4 selecting which LEDs will be lit for each count. While the clock is feeding the counter (i.e., S1 is closed), all seven LEDs will appear to be continuously lit, of course. The parts list for this project is given in Table 17-3.

Figure 17-7 shows how the LEDs should be mounted for traditional dice face patterns. Follow the numbering carefully, or you'll get some odd patterns. The count will be correct, but the

235

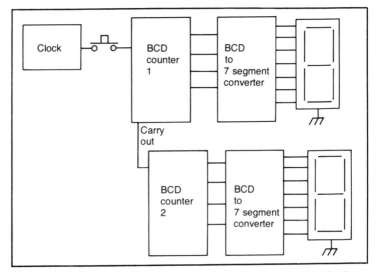

Fig. 17-4. Two-digit decimal numbers may be generated by using this circuitry.

display will not resemble a die. The LEDs to be lit for each count are indicated in Table 17-4 and Fig. 17-8. Since many games call for a pair of dice, you might want to build two of these circuits in a single case with a common power supply.

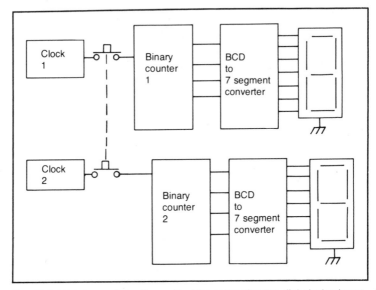

Fig. 17-5. Here is an alternative method for generating two-digit decimal numbers.

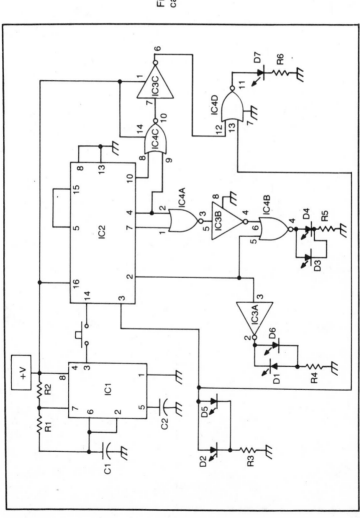

Fig. 17-6. An electronic dice circuit can be used in many games.

IC1	555 timer
IC2	CD4017 decade counter
IC3	CD4049 hex inverter
IC4	CD4001 quad NOR gate
D1 - D6	LED
R1, R2	1 k resistor
R3 - R6	330 ohm resistor
C1, C2	0.01 μF disc capacitor
S1	SPST (normally open pushbutton switch)

ROULETTE WHEEL

Another popular game project that can be built around a random number generator is a roulette wheel. The players try to guess

Table 17-4. The LEDs that Are Lit for Every Possible Output of the Electronic Dice Project of Fig. 17-6 Are Outlined Here. See Figs. 17-7 and 17-8.

COUNTER OUTPUT	LEDS LIT (SEE FIGURE 17-8).
	1 2 3 4 5 6 7
1	- - - - - - x
2	x - - - - x -
3	x - - - - x x
4	x - x x - x -
5	x - x x - x x
6 (0)	x x x x x x -

x = LED on
- = LED off

which LED will be on when the counter stops.

A simple circuit for a roulette wheel is shown in Fig. 17-9. This is simply a binary random number generator driving a 74154 4 to 16

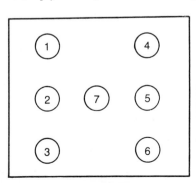

Fig. 17-7. The LEDs for the electronic dice project of Fig. 17-6 should be arranged in the traditional die face pattern.

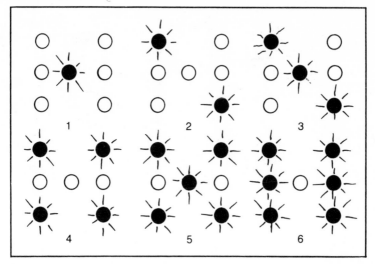

Fig. 17-8. Each possible output count from one to six produces an unique pattern of activated LEDs.

line decoder. For any given count only a single LED will be lit. While S1 is closed, of course, all 16 LEDs will appear to be constantly lit. The parts list for this circuit is shown in Table 17-5. While functional, this circuit leaves quite a bit to be desired. The spinning wheel effect is not at all realistic, and that seems to remove a lot of the excitement from the game.

A better roulette wheel circuit is shown in Fig. 17-10, with the parts list shown in Table 17-6. Here, the 555 timer based clock is replaced with a clock formed from three sections of a 7404 hex inverter. When pushbutton switch S1 is momentarily closed, the circuit will start oscillating. The frequency will start out fairly high (the frequency can be changed by changing the value of capacitor C1), but it will gradually slow down and stop at a rate determined by

Table 17-5. Parts List for the Simple Electronic Roulette Wheel Circuit of Fig. 17-9.

IC1	555 timer
IC2	74175 quad D flip-flop
IC3	74154 4 to 16 line decoder
D1 - D16	LED
R1, R2	100 k resistor
R3 - R18	330 ohm resistor
C1	0.1 μF disc capacitor
C2	0.01 μF disc capacitor
S1	SPST (normally open pushbutton switch)

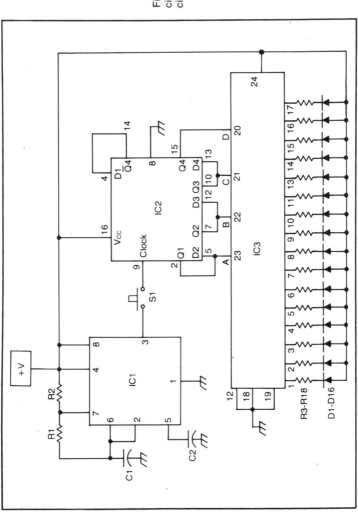

Fig. 17-9. This simple roulette wheel circuit is functional, but not too exciting.

capacitor C2. Increasing the capacitance of this component will cause the "wheel" to "spin" longer. A smaller capacitor will result in a shorter sequence.

As a bonus, a clicking sound effect is added each time the count is incremented. If you don't want the sound effect you can eliminate IC1D, Q1, R4, C3, and the speaker from the circuit. Most garden variety NPN transistors will work here. The specifications are not at all critical. Of course, to give the roulette wheel effect, the 16 LEDs should be arranged in sequence in a circle. As each LED is lit and extinguished in turn, the effect of a spinning wheel will be given.

In either of these circuits the clock frequency should be lower than that used in the random number generator circuits discussed earlier. We want to see the individual LEDs go on and off in turn to give the moving dot and spinning wheel effect.

In the circuit of Fig. 17-9 the wheel will "spin" at a fixed rate (approximately 48 Hz for the component values shown). The frequency determining components are R1, R2, and C1. The other circuit (Fig. 17-10) will start out with a fast "spin" which will gradually slow down and stop, much like a real roulette wheel.

CLOCKS

Counter circuits can also be used to build digital clocks and timers. This time we mean clock in the usual sense—a device that tells you what time of day it is. The first thing you need when designing an electronic timepiece is an accurate timebase. This is basically the same as the clock oscillators used to keep various

**Table 17-6. The Improved Electronic Roulette
Wheel Project of Fig. 17-10 Can Be Constructed Using These Components.**

IC1	7404 hex inverter
IC2	74175 quad D type Flip-flop
IC3	74154 4 to 16 line decoder
Q1	almost any low power NPN transistor
D1	1N4148 diode (or similar)
D2 - D17	LED
R1	120 k resistor
R2	470 k resistor
R3, R4	10 k resistor
R5 - R20	330 ohm resistor
C1, C3	2.2 μF 25 volt electrolytic capacitor
C2	330 μF 25 volt electrolytic capacitor
S1	SPST (normally open pushbutton switch)
SPKR	small speaker

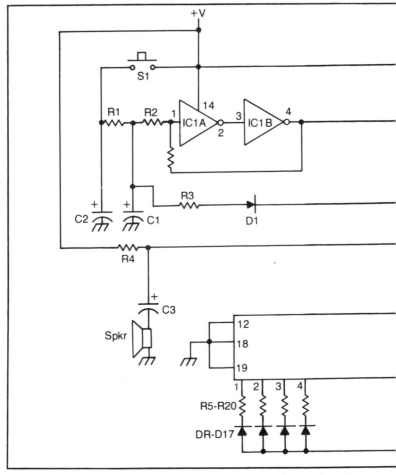

Fig. 17-10. This circuit can be used to create a more realistic and exciting electronic roulette wheel.

digital circuits in synchronization. A digital time-telling clock essentially counts the numbers of seconds, minutes, and hours. Of course, it needs some way to know just how long a second is so it can be counted. The timebase generates a signal with a precise frequency, so that the circuitry can count X number of pulses per second.

This timebase frequency must be extremely accurate. It might seem that there is little difference between .95 second and 1 second. But when you multiply that 5% error over a 24-hour day, you end up with a 22-hour, 48-minute day! That's not very good

242

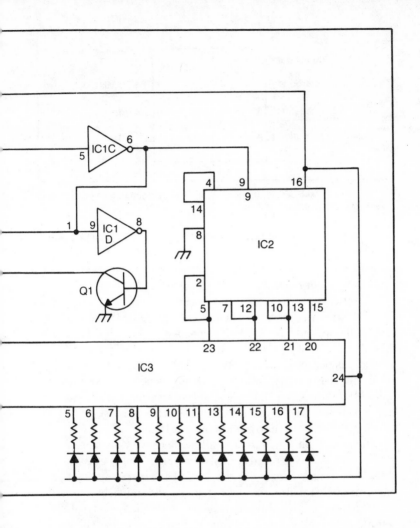

timekeeping. About the only thing that kind of clock would be useful for is paperweight duty!

Most electronic clock circuits work with a timebase of 60 Hz (60 pulses per second). This tradition stems from the ac electric power lines which operate at a 60 Hz rate. An electrically powered clock can use the ac power source itself as a timebase. Adapting the ac power signal for use in a digital circuit is difficult, and with modern technology, it's more trouble than it's worth.

Most digital clocks work with some sort of crystal oscillators. Quartz crystals make very precisely held frequencies possible.

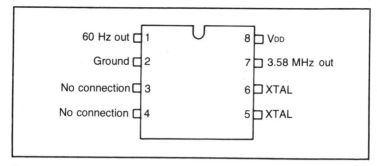

Fig. 17-11. The MM5369 is an eight-pin IC that can generate a precise and reliable 60-hertz timebase signal.

Incidentally, when digital watches first became popular, many were proudly touted in the ads as being quartz controlled. *All* digital watches are quartz controlled. A crystal oscillator is used for the timebase, and crystals are slabs of quartz.

Most crystals operate at frequencies much higher than 60 Hz. Generally, their resonant frequencies are above 1 MHz (1,000,000 Hz). Additional counter stages are needed to drop this frequency down. The MM5369 (shown in Fig. 17-11) is a specially designed IC for just this purpose. It generates an extremely precise and stable 60 Hz timebase signal from a 3.579545 (generally shortened to 3.58) MHz (3,579,545 Hz) crystal. This particular input frequency was selected, since it is the frequency used in color burst oscillators in color TV sets, so they are readily available.

A single 8-pin chip provides all the necessary division. Actually, except for the V_{DD} and ground connections, only three pins are used. Two connections go to the 3.58 MHz crystal, and one provides the 60 Hz timebase output signal. In case it might be needed in some circuits, pin #7 also provides a 3.58 MHz output signal. The remaining two pins are not internally connected to the chip.

Table 17-7. Only a Few Parts Are Needed to Generate
a Reliable and Accurate 1-Hz. Signal, Using the Circuit Shown in Fig. 17-12.

IC1	MM5369 60 Hz timebase
IC2, IC3	CD4017 decade counter
	(IC2 = + 6, IC3 = + 10)
R1, R2	10 megohm resistor
R3	1 k resistor
C1	*** see text
C2	*** see text
XTAL	3.58 MHz colorburst crystal

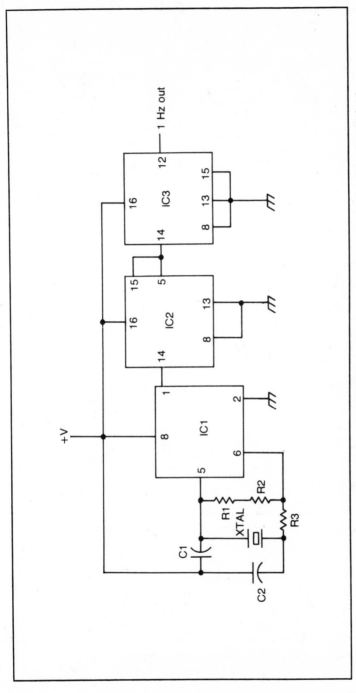

Fig. 17-12. By adding a couple of divider/counters to the MM5369 timebase, an accurate one-hertz signal can be obtained.

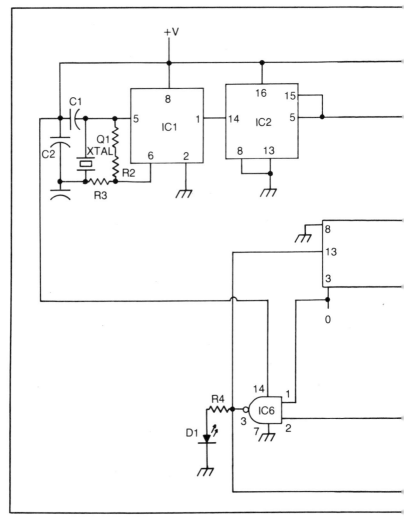

Fig. 17-13. This timer circuit can measure a 60-second period.

The 60 Hz timebase signal must be divided by 60 to get a once per second signal that can be counted by the clock. Figure 17-12 shows a practical circuit for a 1 Hz output signal. Notice the three resistors and two capacitors in the circuit with the crystal itself. They are there to improve stability. According to the manufacturer, C1 should have a value of 6.36 pF, and C2 should be 30 pF. Unfortunately, these values are not commonly available. C1 could be 10 pF and C2 may be 47 pF without significantly hurting the

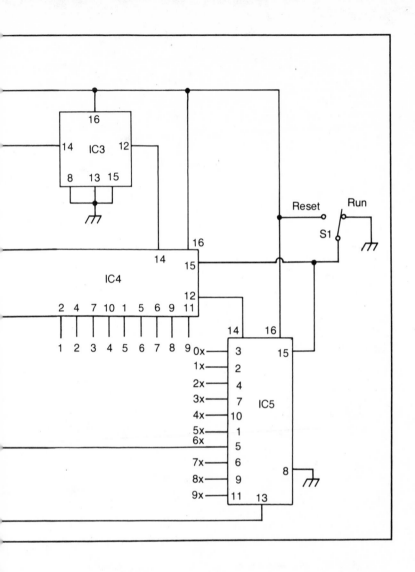

precision of the output frequency. If you want perfection, trimmer (variable) capacitors may be included in the circuit for fine tuning.

The 60 Hz timebase signal is fed to IC2, a CD4017 IC connected as a 6-step counter. This drops the frequency to 10 Hz. IC3, and other CD4017, divides the signal frequency by an additional factor of 10, resulting in a precise 1 Hz output. The complete parts list for this circuit is given in Table 17-7.

Adding a pair of additional timers and an AND gate, along with

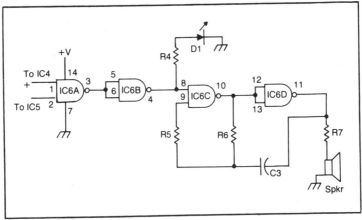

Fig. 17-14. Adding these components to the 60-second timer circuit of Fig. 17-13 will cause a tone to be sounded when the circuit times out.

a handful of other components, makes a 60 second timer. When S1 is moved to the RUN position, sixty seconds will be counted, then the LED will go on, until S1 is moved to the RESET position. The circuit is shown in Fig. 17-13, and the parts list is given in Table 17-8.

By connecting different outputs from IC4 and IC5 to the inputs of the AND gate, you can select a timing cycle of anywhere from 1 to 99 seconds (in one second intervals). A pair of ten position thumbwheel switches can turn this project into a fine programmable timer.

Another useful modification to this project is shown in Fig. 17-14 (parts list in Table 17-9). By replacing the AND gate (IC6 in Fig. 17-13) with this circuit (using a single quad NAND gate package and a handful of discrete components), when the timing period is over, the LED will light, and a tone will be sounded, until the circuit is reset. IC6A and IC6B behave like the original AND gate, while

Table 17-8. This Is the Parts List for the 60-Second Timer Circuit of Fig. 17-13.

IC1 - IC3	see Table 17-7
R1 - R3	see Table 17-7
C1, C2	see Table 17-7
XTAL	see Table 17-7
IC4, IC5	CD4017 decade counter
IC6	CD4081 quad AND gate
D1	LED
R4	330 ohm resistor
S1	SPDT switch

Table 17-9. These Components Can Be Added to the 60-Second Timer Circuit of Fig. 17-13 to Add an Audible Alarm. The Circuitry Is Shown in Fig. 17-14.

Eliminate IC6, R4, and D1 in Fig. 17-13, and replace with:

IC6	CD4011 quad NAND gate
D1	LED
R4	330 ohm resistor
R5	1 megohm resistor
R6	100 k resistor
R7	100 ohm resistor
C3	0.01 μF disc capacitor

IC6C and IC6D are connected as a tone generator with an output of approximately 1 kHz (1000 Hz).

Back-tracking somewhat, connecting the 1 Hz source circuit of Fig. 17-12 to the input of the circuit illustrated in Fig. 17-15, will provide a clock with minute readouts from 00 to 59. Once the count reaches 60, the counters are reset to 00. The count is displayed on a pair of seven-segment LED displays. The parts list for this subcircuit is given in Table 17-10.

Figure 17-16 shows the circuitry for adding the display for hours (ranging from 01 to 12). IC10 and IC11 count each group of sixty minutes. Alternatively, the signal from the minutes counter reset (from IC7C and D) could be used to trigger the hours counter.

The ones digit is wired in the usual fashion, but the tens column can use a few short cuts. This digit will always be either 0 or 1. Segments b and c will always be lit, so they are tied directly to the positive power supply (through current dropping resistors). If the count is less than 10, then segments a, d, e, and f are lit to produce a 0. Segment g is never lit. The parts list for the hours display circuit is given in Table 17-11. The complete block diagram for the clock project is shown in Fig. 17-17.

Table 17-10. The Minutes Portion of a Practical Digital Clock Can Be Constructed Using These Components Wired as Shown in Fig. 17-15.

IC4, IC5	CD4017 decade counter
IC6	CD4518 dual BCD counter
IC7	CD4011 quad NAND gate
IC8, IC9	CD4511 BCD to 7 segment decoder
DIS1, DIS2	seven segment LED display, common cathode
R4, R7	330 ohm resistor

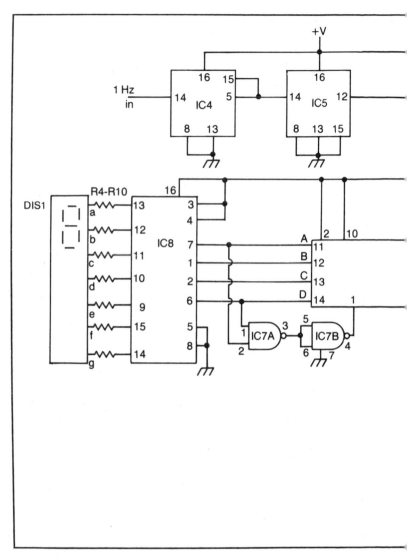

Fig. 17-15. This circuit can display the minutes counter in a digital clock.

COUNT-DOWN TIMER

Another useful timing circuit would be a timer that counts backwards. A single-digit count-down timer is shown in Figs. 17-18 and 17-19.

Figure 17-18 is the manual input for entering any beginning number from 1 to 9 in the proper BCD format. The actual count-

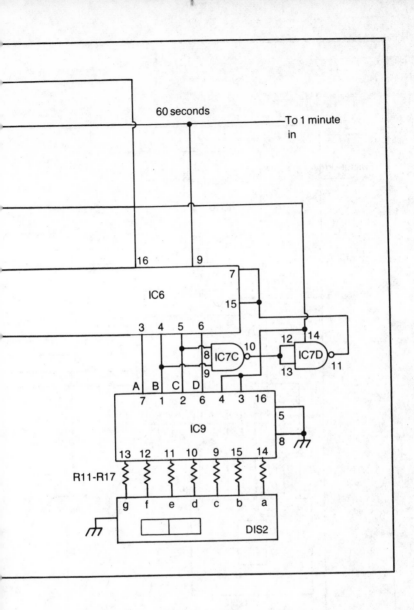

down circuit is shown in Fig. 17-19. The input signal to be counted can come from the 1 Hz signal source from Fig. 17-12. By adding another divide-by-6 and divide-by-10 counter the circuit can count down 1 to 9 minutes. Additional stages (for multidigit countdowns) can be added by tapping off the borrow out signal from 74193 up/down counter. The parts list is given in Table 17-12.

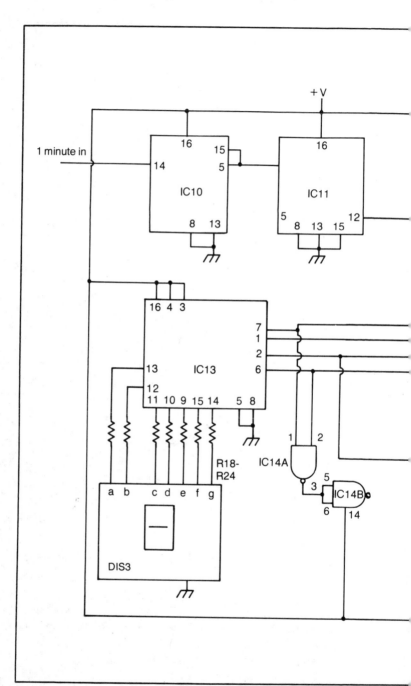

Fig. 17-16. The hours portion of a digital clock is controlled using this circuit.

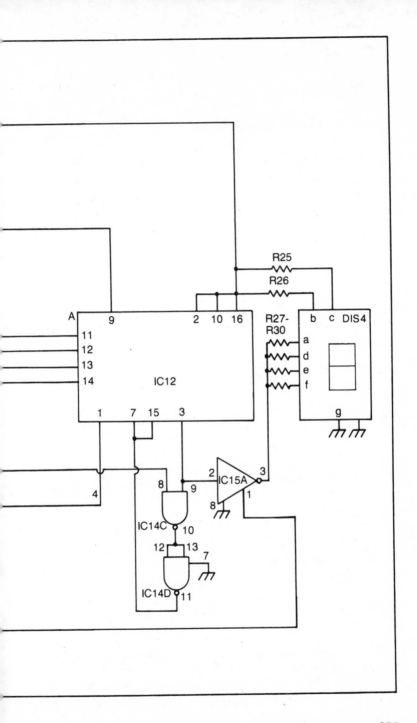

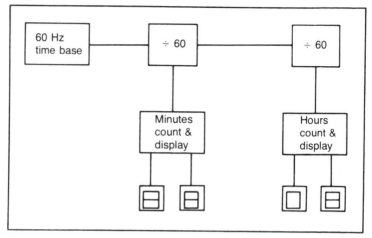

Fig. 17-17. The complete clock project is shown here in block diagram form.

Table 17-11. Here Is a List of the Parts Needed
to Add the Hours Display Circuitry Illustrated in Fig. 17-16.

IC10, IC11	CD4017 decade counter
IC12	CD4518 dual BCD counter
IC13	CD4511 BCD to 7 segment decoder
IC14	CD4011 quad NAND gate
IC15	CD4049 hex inverter
DIS3, DIS4	seven segment LED display, common cathode
R18 - R30	330 ohm resistor

Table 17-12. These Parts Are Used in the
Count-Down Timer Project of Figs. 17-18 and 17-19.

IC1	7404 hex inverter
IC2	74193 up/down counter
IC3	7446 BCD to 7 segment decoder
DIS1	7 segment LED display, common anode
R1, R2	1 k resistor
R3 - R9	330 ohm resistor
S1	4 pole, 9 throw rotary switch
S2	SPST (normally open pushbutton switch)

Countless other counter projects are also possible, but I feel we've covered a pretty good cross section in the last two chapters. Now we will move on to some other projects. Some of these will include counters too. Counters are so basic to the whole idea of digital electronics, they fit into a vast number of different projects.

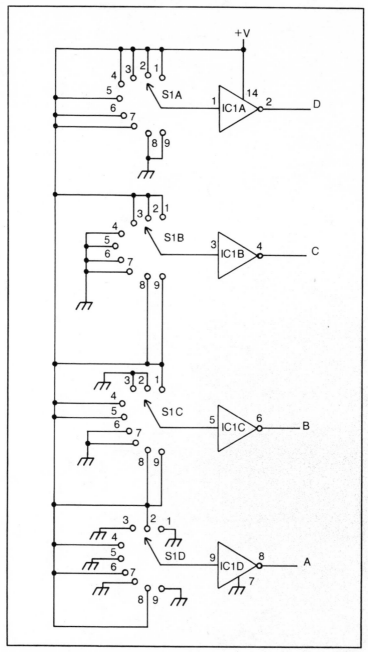

Fig. 17-18. This switching circuit is used to select the beginning count for the count-down timer circuit of Fig. 17-19.

255

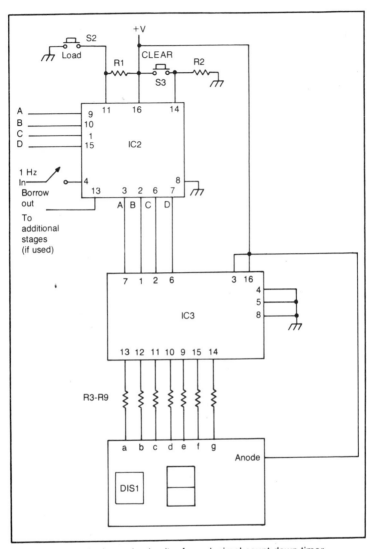

Fig. 17-19. Here is the main circuitry for a decimal count-down timer.

Chapter 18

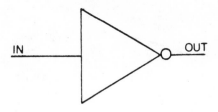

Digital Voltmeters and Multimeters

The multimeter has always been the electronics technicians "right arm." The two basic analog multimeters are the VOM (volt-ohm-milliamp meter) and the VTVM (vacuum-tube voltmeter). Semiconductor equivalents to the VTVM have also been developed. The quantity to be measured on either a VOM or a VTVM is displayed on the face of an analog meter, such as the one shown in Fig. 18-1.

As useful as these devices are in many measurement situations, they do suffer some deficiencies in certain cases. It can be difficult to get a precise reading, especially when the pointer is in a crowded portion of the scale. Being mechanical devices, analog meters may get out of adjustment, especially if the unit has been physically jolted or dropped. For instance, if the pointer needle is slightly bent, the readings obtained will be off. Analog meters are also subject to overshoot (the pointer moves too fast, and goes past the correct reading) and undershoot (the same problem, except the pointer is moving in the opposite direction—from a high value to a low one).

If too high a voltage is fed to an analog meter, the pointer can be slammed heavily against the far edge of the scale. This could cause the needle to be bent, or cause other damage. The analog meter movement is relatively delicate, and can be easily damaged by physical abuse including accidents and overvoltages.

By using modern electronics technology, and especially digital devices, many of these problems will be overcome. This has led to

257

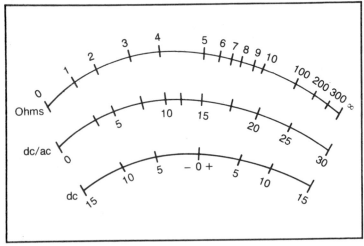

Fig. 18-1. The face of an analog VOM usually has several scales.

the growing popularity of DMMs, or digital multimeters. Since the output is presented directly in numerical form, there is no ambiguity when reading values. Also, since there are no mechanical, moving parts problems like overshoot, undershoot, and slamming, just don't exist.

A digital meter is not perfect for all applications, however. In some cases, the analog meter still offers superiority. For example, in any measurement of changing values, where you need to monitor the trends of changes, such as a charging capacitive circuit, a digital readout would be confusing, at best. The movement of the pointer across the face of the analog meter would clearly show the trends exhibited by the measured value. For instance, the voltage across a charging capacitor would slowly rise to a maximum value, and then stop. When the capacitor is discharged, the reverse would occur. Many technicians and hobbyists today keep both an analog meter and a digital meter around, so they will have the best equipment for either type of job.

READOUTS

Most digital meters use either seven-segment LED (light-emitting diode) or LCD (liquid-crystal display) displays. We have already worked with the seven-segment LED display in earlier chapters. Before discussing digital readouts in detail, let's examine a simpler type of readout, which is a sort of cross between analog and digital. This is the dot or bargraph.

```
1 2 3 4 5 6 7 8 9 10
OOOOOOOOOO
```

Fig. 18-2. Separate LEDs can be combined to create a dot or bargraph display.

Dot and Bargraph Displays

Dot and bargraph displays are made up of a series of separate LEDs, usually arranged in a line, as illustrated in Fig. 18-2. There is no law that says they have to be set up in a row. Sometimes its more convenient to use another arrangement, such as the one shown in Fig. 18-3.

In a dot display, only a single LED is lit at a time. As an example, let's assume we have a simple voltmeter with a five LED

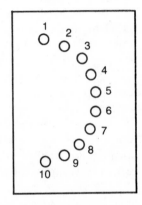

Fig. 18-3. The LEDs in a dot or bargraph display may be arranged in any convenient pattern.

dot display readout. Each LED indicates one volt. If LED 3 is lit, as in Fig. 18-4, we know the applied voltage is greater than three, but less than four volts.

Bargraph displays are quite similar, except all of the LEDs below the measured value are simultaneously lit, creating a line effect. The longer the line, the larger the input signal. Figure 18-5 shows a bargraph display that is the equivalent of the dot display shown in Fig. 18-4. Both devices are displaying the same input value.

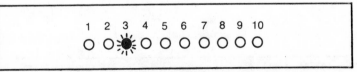

Fig. 18-4. In a dot display only a single LED is lit at any given time.

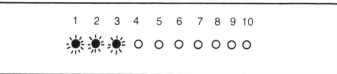

Fig. 18-5. In a bargraph display all LEDs below the measured value are lit.

Often, the separate LEDs in a bargraph display will be placed in a single housing with a translucent lens to give more the effect of a solid line, than a row of dots. The difference is purely cosmetic. See Fig. 18-6.

Clearly dot and bargraph displays are not suitable for precise measurements. There is no way to distinguish between values that fall between adjacent LEDs. In our examples, 3.0 volts, 3.6 volts, and 3.25 volts will all produce the same display readout. This problem could conceivably be improved by adding more LEDs, but this solution is severely limited. If too many LEDs are included in the display, it will be difficult to interpret readings quickly.

The biggest advantage of dot and bargraph displays is that they can be read at a glance. The readings are not precise, but in many applications only approximate readings are demanded anyway. And dot or bargraph displays are significantly lower in cost than numerical readouts. Changing trends can very easily be seen on these displays. Relatively brief pulses can also be spotted far easier than with a mechanical analog meter.

Because an analog meter responds to changing signals by mechanically moving a pointer, its reaction time is fairly slow. Friction and mechanical inertia tend to hold it back. A short pulse might not show up on a mechanical meter at all. Or if it does produce a visible response, it probably won't show the correct peak voltage. For instance, assume we are measuring a 1-volt signal. Suddenly, a 5-volt pulse appears. The meter's needle begins to swing up the scale. But by the time it reaches the 3-volt position, the pulse may be gone, and the input signal has returned to its original 1-volt steady state. The needle will move back down to the 1-volt position,

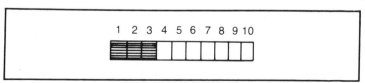

Fig. 18-6. Many dedicated bargraph units with a common Housing and LENs are available.

without ever displaying the true 5-volt peak. According to this hypothetical meter, the peak voltage is only 3 volts.

If the pulse is a clean square wave with very fast rise and fall times, the needles might overshoot the one-volt position. This meter's response, and the actual signal are compared in Fig. 18-7. The rather striking differences between these two signals is obvious, and could mean the difference between successfully understanding what's happening in the circuit, and sending the technician off on a confusing (and potentially costly and time-consuming) wild goose chase.

An LED dot or bargraph display has no mechanical moving parts, so it can respond to changing signals far faster than any analog meter possibly could. What little delay there is, is in the microsecond range. A microsecond is one millionth (0.000001) of a second. Believe me, you're not going to notice any delay. Problems like overshoot and undershoot also don't occur. Since a brief peak pulse is displayed as a quick flash of light from the LEDs, persistence of vision allows you to monitor pulses that are faster than you would ordinarily be able to distinguish.

A simple resistance ladder bargraph circuit is shown in Fig. 18-8. The resistors act as a multiple voltage divider. Their values are selected to cause each LED to turn on at the desired voltage. If the voltage applied to this circuit is such that the voltage at point B

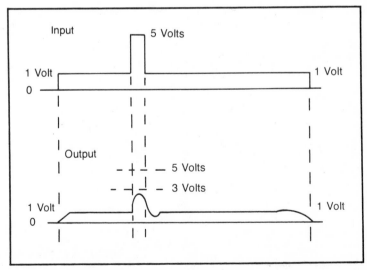

Fig. 18-7. These input and output signals indicate some of the shortcomings of a mechanical analog meter.

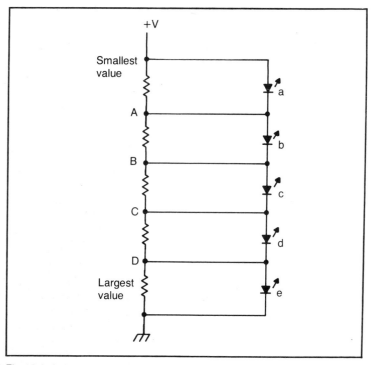

Fig. 18-8. A simple resistance ladder network can be used as a bargraph display driver.

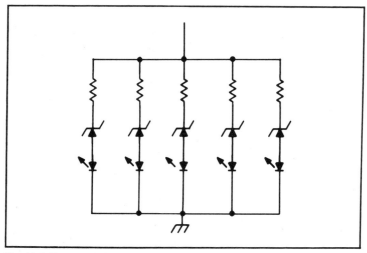

Fig. 18-9. Zener diodes can also determine the switching points for the individual LEDs in a bargraph display.

is just enough to turn on an LED, LEDs c, d, and e will be lit, while LEDs a and b will remain dark.

Another simple LED bargraph circuit is illustrated in Fig. 18-9. This time, zener diodes are used to determine the switch on voltages for each of the LEDs. If a given zener diode's breakdown voltage is exceeded, the associated LED will be lit.

Both of these circuits are essentially passive devices. They can load down the circuit being monitored in some cases. This could cause false readings. Isolating the display driver circuit from the circuit under test with a high input impedance device such as an op amp IC can cut down such potential problems considerably. In fact, an op amp input buffer stage will exhibit a much higher input impedance than most analog VOMs.

A high input impedance active bargraph display driver circuit is illustrated in Fig. 18-10. Almost any low-power NPN transistor will work in this application. Transistors Q1 and Q2 are connected as a Darlington pair. The network comprised of Q_x, R_a, R_b, and an LED is repeated for each stage (for each LED you want in the output display).

Most modern bargraph display drivers work on the comparator principle. A separate comparator is used for each stage (i.e., each

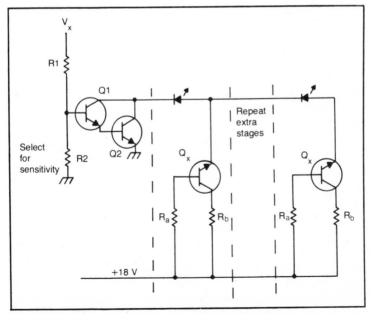

Fig. 18-10. This circuit is a basic active bargraph driver circuit.

LED in the output display). Integrated circuits containing four comparator units in a single package are available, and frequently used in this manner. A typical circuit is shown in Fig. 18-11.

A resistor voltage divider network is extended from + Vcc to ground has several tap-off points that are fed to the B inputs of each comparator. The A input of all of the comparators is the unknown voltage. If the unknown voltage (V_x) exceeds the reference voltage for a given stage (that is, if A > B), the output of that comparator goes high, and the associated LED is turned on. All LEDs representing voltages lower than V_x will be turned on.

A number of more advanced dot and bargraph display driver circuits are available in IC form. For instance, the LM3914, and the LM3915 are ten-stage LED dot/bargraph display drivers. Each chip contains a reference voltage source, a voltage divider network and ten individual comparators. The pinout diagram for the LM3914 is shown in Fig. 18-12. The on-chip voltage reference source is preci-

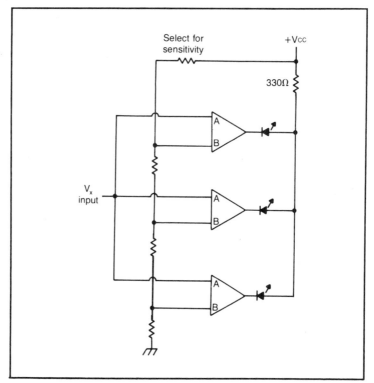

Fig. 18-11. Most bargraph circuits use a number of comparators to control the LEDs.

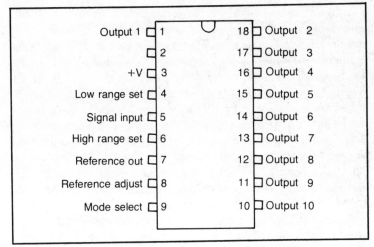

Fig. 18-12. The LM3914 is a dedicated bargraph display driver IC.

sion regulated and will hold its value constant with supply voltages ranging from 3 to 25 volts.

Only the LEDs themselves, a pair of resistors, and an electrolytic capacitor are required to make a complete bargraph display voltmeter around the LM3914 IC. The circuit is shown in Fig. 18-13. The voltage to be measured is applied to pin #5.

Pin #9 (mode) calls for some comment. If this pin is left unconnected, as shown in the schematic diagram, the LM3914 will operate in the bargraph mode. All LEDs below the current input voltage will be simultaneously lit. Refer back to Fig. 18-5. On the other hand, connecting pin #9 to the common anodes of the display LEDs puts the unit into the dot mode. Only a single LED is lit at any one time. Refer back to Fig. 18-4.

Notice that the display LEDs are hooked up in the common anode format, and require a positive voltage source. This may be the same as the main power supply in some applications, or it may be a separate power source. The 10 μF electrolytic capacitor functions as a bypass capacitor to prevent parasitic oscillations.

The measurement range of this simple voltmeter can be set by applying the appropriate voltages to pins #4 and 6. Pin #4 sets the bottom end of the measurement range, and pin #6 determines the maximum measured voltage. If the circuit is wired as shown in Fig. 18-13, the range will be from 0 to 2.5 volts. Each LED will represent an increment of 0.25. For example, if the applied voltage is 1.75 volt, the first seven LEDs will be lit.

Mounting and soldering the separate LEDs can be a bit of a nuisance, so the NSM series of modules include the driver circuitry and the LED display on a small factory-built PC board. Twelve edge terminals are arranged along one side of the board, as illustrated in the pinout diagram for the NSM3915, which is given in Fig. 18-14. The function of each of these pins is identified in Table 18-1.

Essentially this unit works in the same way as the LM3914 chip we have just discussed, except for the convenience of a one-piece driver/display unit. A few external resistors and capacitors are required. A typical circuit is shown in Fig. 18-15. This circuit operates over a logarithmic, rather than linear range, with each successive LED being activated at 3 dB intervals. The NMS driver/display units measure a mere 1.99 inches by 0.85 inches.

Dot and bargraph displays are suitable for applications where rapidly changing values that must remain within a certain range must be monitored, without any need to determine precise intermediate values. A typical application would be a VU (volume unit) monitor in a sound or recording system. If the signal level is too

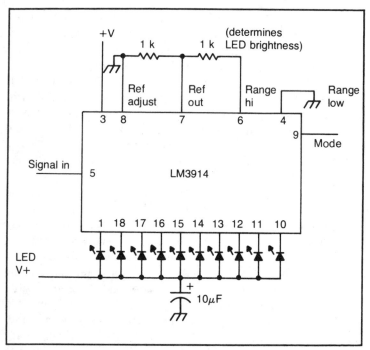

Fig. 18-13. The LM3914 bargraph display driver IC can be used as the heart of a simple bargraph voltmeter.

Table 18-1. The Functions of the Pins on
the NSM3915 Are Outlined Here (See Fig. 18-14).

Pin #	Title	Function
1	V_{LED}	Positive voltage for common anode connection of the display LEDs
2	LED 1	Output to external device, controlled along with LED 1 of the on-board display
3	Ground	Circuit ground
4	V+	Positive power supply for the driver circuitry
5	R_{LO}	The low end of the measurement range is set via a voltage applied to this pin
6	Signal In	Input for the voltage to be measured
7	R_{HI}	Set high end of measured range
8	Ref	Output carrying the internal reference voltage
9	Ref Adj	Control the reference voltage
10	Mode	Dot or Bargraph display select
11	LED 9	External output in parallel with LED 9
12	LED 10	External output in parallel with LED 10

low, the desired signal may be lost in system noise. On the other hand, if the signal level is too high, distortion may result. Music and other sounds are constantly changing in volume at a fairly rapid rate. The exact level is usually unimportant as long as it is within the acceptable range. This is very easy to see on a dot or bargraph display. This is especially true if different colored LEDs are used—for example, green for acceptable, and red for under- or over-range.

Digital Displays

When precise measurements are required, a true digital readout is an excellent choice. A display that says 5.67 volts, for example, is quite unambiguous. Of course, the number of digits limits the resolution of the readout. A three-digit display would read 4.32 for measurements of 4.318, 4.3209, or 4.323.

Practical meters also have a certain amount of inherent error, usually expressed as ± x% of the full-scale reading. For most com-

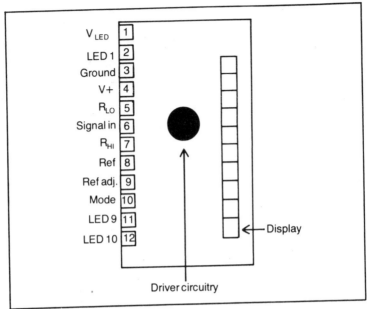

Fig. 18-14. The NSM3915 consists of a dot/bargraph display driver, and the LEDs themselves.

mercial digital voltmeters, this is ±.1 or .2%, if not even lower. This is usually not a serious amount of error. For instance, let's assume we have a meter with a maximum readout of 999.9, and a

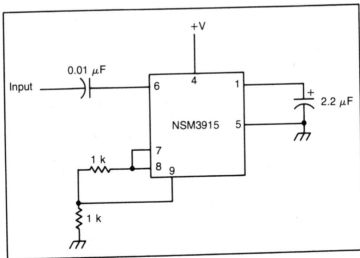

Fig. 18-15. This is the basic circuit for a NSM3915 voltmeter.

268

±.1% error. If the measured voltage is 241.5 volts, the readout might display anything from 240.5 to 242.5. This is a relatively small error, especially when compared to analog meters, where a full scale error of ±1% is considered exceptionally good.

Most digital voltmeter circuits have a ± one-digit error. That is, a reading of 41.7 could be indicative of a measurement (ignoring other errors) of 41.6, or 41.8. Since this error shows up in the least significant digit, it is rarely of significance.

The bobbing least significant digit occurs because of the way voltages are measured in a digital meter. The measured voltage is used to charge and/or discharge a capacitor. The time this takes is measured by counting the pulses from a known frequency source. On some counting cycles, a partial pulse may be included incrementing or decrementing the count by one.

Digital meters are usually classified by the number of digits they have. A range selector switch moves the decimal point to extend the range of the meter. A three-digit readout meter could have four basic ranges with maximum counts of 999, 99.9, 9.99, and .999.

Many digital meters also include what is called a half-digit. This is actually an overflow digit. If the basic range is exceeded, the half-digit displays a 1, otherwise it is a 0. Where a three-digit display has a maximum readout of 999, a 3½-digit display can indicate values up to 1999. The range is essentially doubled. Displays of 2000 or more, however, are not possible. The most significant digit can be only a 0 or a 1.

Digital meters are made up of seven-segment displays, with the addition of decimal points. Many of these displays also feature a negative sign for reverse polarity signals. Additional LEDs may be used to indicate the selected range, battery condition and overflow.

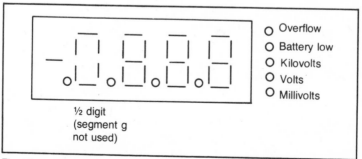

½ digit
(segment g
not used)

Fig. 18-16. A typical digital meter has a readout display that looks something like this.

A typical digital meter display is shown in Fig. 18-16.

While many digital meters use LEDs, some use LCDs, or liquid-crystal displays. A LCD consists of a special liquid sandwiched between two pieces of glass or plastic. When an appropriate electrical charge is placed on a segment, it turns opaque. The segments are arranged in the same manner as LED seven-segment displays.

LCDs do not emit light themselves, so they may be difficult to see under poor lighting conditions. On the other hand, LED displays may appear to wash out under bright lighting. They also consume far more current than comparable LCD units. Different drive circuitry is required for LCD and LED display units, but the principles are functionally identical.

DIGITAL VOLTMETER CIRCUITS

Most digital voltmeters function by converting the voltage to be measured to time, by using it to charge and/or discharge a capacitor. This time period is then measured by counting the number of pulses from a known precision frequency source that occur during the time period.

As simple as this basic scheme sounds, there are some complications involved in putting it into practice. The biggest problem is that an applied voltage charges a capacitor exponentially (see Fig. 18-17), but we want the readout to display a linear relationship to

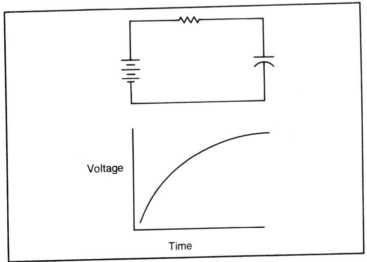

Fig. 18-17. A voltage charges a capacitor exponentially.

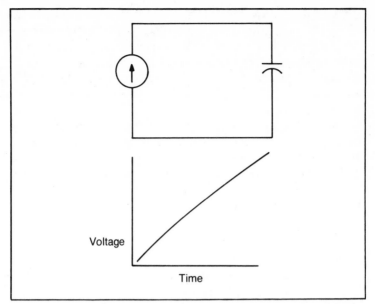

Fig. 18-18. A constant current source can be used to charge a capacitor linearly.

the input voltage. That is, if a one-volt input produces a count of 1000, for instance, than a two-volt input should result in a count of 2000, a 0.5-volt input should result in a count of 500, and so forth. A capacitor may be charged linearly if the voltage is replaced with a constant current source, as shown in Fig. 18-18. This is what is done in most digital voltmeters.

Essentially, this process of converting the input voltage to be measured to a counter output is an analog to digital conversion. The number of times the conversion, or counting cycle takes place varies. In most practical digital meters approximately three to five complete counting operations are done each second. If less than three conversions per second are performed, the instrument will tend to have an objectionably slow response to changes in the input voltage. A conversion rate faster than about 5 times per second would tend to cause the display readout to jitter and become difficult to read. There are two basic approaches to the analog to digital conversion process that are used in digital voltmeters. They are called single-slope conversion and dual-slope conversion.

Single-Slope Conversion

Single-slope conversion is the simpler of the two basic

methods of analog to digital conversion used in digital voltmeter circuits. It is rarely used in modern commercial units for reasons to be explained shortly, but it is a functional approach to the task of converting the unknown input voltage to a linear time period. A simplified single-slope conversion circuit is shown in Fig. 18-19.

A constant source is used to linearly charge capacitor C. The charge on this capacitor is used as a reference voltage that is fed into the comparator, along with the unknown input voltage to be measured. When S1 is closed, the capacitor is bypassed to ground, allowing it to discharge, resetting the reference signal. This means the reference voltage starts out at zero, and linearly rises to its maximum level. As long as the input voltage is greater than the reference voltage, the output of the comparator will be a 1. As soon as the reference voltage passes the level of the unknown input voltage, the comparator output switches to a logic 0. These signals are illustrated in Fig. 18-20.

Following the comparator stage, there is a three input AND gate. As you know, the output of an AND gate is a logic 1 if and only if all three inputs are at the logic 1 level. If one or more of the inputs is fed a logic 0, the output will be a logic 0.

S2 is normally in the logic 1, "COUNT" position. When the counter is being reset, this switch is grounded to produce a logic 0 input to the AND gate, inhibiting the oscillator pulses to the counter.

Assuming S2 is in its normal, logic 1 position, the output of the

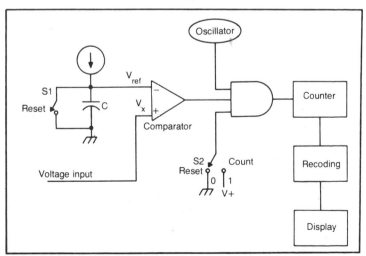

Fig. 18-19. This is a simplified single slope conversion circuit.

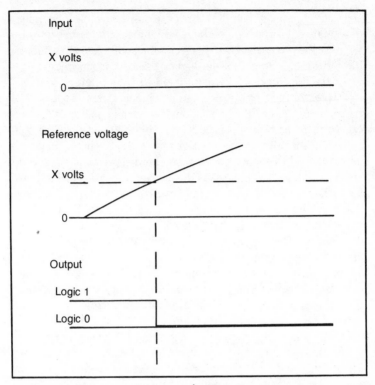

Fig. 18-20. The input and output signals for the comparator stage demonstrate how the single-slope converter works.

AND gate is controlled by the comparator. During the capacitor charging/measurement cycle the AND gate passes the oscillator pulses to the counter for the time it takes the reference voltage charging the capacitor to exceed the unknown input voltage. Then the comparator output drops to logic 0, inhibiting the AND gate and preventing further oscillator pulses from reaching the counter. The pulses that are allowed to pass through the gate are fed into the counter stage. The result of the counting operation is decoded and displayed.

The charging rate of the reference capacitor and the oscillator frequency are set up so that the displayed count is directly proportional to the input voltage being measured. For example, a count of 1000 pulses would produce a display of 10.00 for an input of ten volts.

The operation of the single-slope converter is relatively straightforward, but there are some significant shortcomings to this approach. For one thing, the circuit can only recognize input pulses of a single polarity. If the voltage being measured is negative and the reference signal is positive, no counting will take place, because the reference voltage will be higher than the input voltage even when the capacitor is fully discharged (0 volts reference).

Overvoltages can also create problems unless special additional circuitry is added. If the constant current source is setup to charge the capacitor to a maximum level of 20 volts, and an input voltage of 25 volts is fed to the circuit, the reference voltage will never exceed the input voltage, so the comparator's output will never go to logic 0. This means the oscillator signal is continuously fed to the counter circuitry. It will never stop counting. The counter will reach its maximum count, cycle back to 0, and keep right on going. No readable value will ever be displayed.

The single-slope converter is susceptible to any noise appearing in the input signal. This could cause extreme irregularity and bobbing in the display, possible resulting in a completely indecipherable reading. The accuracy of the readings are dependent on the linearity of the capacitor charging rate. Any irregularities could cause errors in the output readings.

The capacitor and the oscillator must be very precise and stable to insure accurate measurements. If the oscillator frequency drifts, or the output of the constant current source drifts, erronous readings will also occur. An important specification is the stability of the differential voltage required to trip the comparator. Another potential source of problems in a single-slope converter circuit is stray capacitances. A glass epoxy PC board with a good ground plane is strongly recommended for such circuits to minimize stray capacitances.

Dual-Slope Conversion

Many of the problems associated with single-slope converters can be overcome by employing a somewhat more complex form of circuitry known as the dual-slope converter. This type of circuitry offers better accuracy than single-slope conversion circuits, especially long-term accuracy. Drift in the output readings is strongly minimized. A simplified dual-slope converter circuit is illustrated in Fig. 18-21. Note the similarity to the single-slope converter circuit shown in Fig. 18-19.

As the name implies, the measurement cycle of a dual-slope conversion circuit consists of two steps. The capacitor is charged linearly by the reference voltage source, then discharged by the

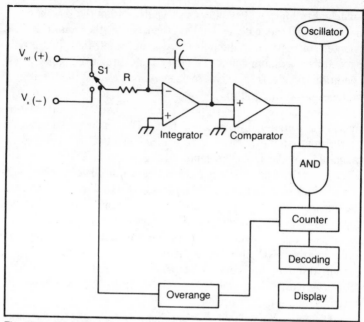

Fig. 18-21. The dual-slope conversion circuit overcomes many of the disadvantages of the single-slope converter.

unknown input voltage. The switching is done automatically by an overrange detector stage that follows the counter stage.

The reference voltage has the opposite polarity as the input voltage to be measured. These voltages are fed to an integrator stage in turn. This circuitry essentially converts the voltage fed to its input to a constant source to linearly charge or discharge the feedback capacitor. The output of the integrator is fed into a comparator stage along with another reference voltage, usually 0 volts (ground).

The measurement cycle begins with the counter stage reset to all zeroes. S1 is automatically set to feed the unknown voltage (V_x) to the integrator, charging the capacitor. Due to the previous measurement cycle, the output of the integrator will start out at a negative value—that is, lower than the comparator reference voltage (0 volts). At this time, the output of the comparator is logic 0, inhibiting the AND gate. This blocks the oscillator pulses from reaching the counter stage. The count is still 0 at this point.

As the input voltage charges the capacitor, the output of the integrator will start to rise linearly. When the integrator's output passes 0, and begins to go positive, the comparator's output is switched to logic 1. This allows the pulses from the oscillator to pass through the gate and be recognized by the counter. The counter starts to count the pulses. At some point, the counter will exceed its maximum count. The overrange circuit recognizes this condition and electronically moves switch S1 to feed the reference voltage to the integrator.

Since the reference voltage is the opposite polarity of the measured voltage, the capacitor begins to discharge. The integrator's output starts out at some positive level determined by the magnitude of the unknown input voltage. The reference voltage discharges the capacitor at a known linear rate. The counter will recycle back to 0 after it was over-ranged by the charging portion of the measurement cycle. It continues to count pulses until the capacitor is discharged enough to drop the integrator's output below 0. This of course, switches the comparator back down to logic 0, inhibiting the AND gate. Since the oscillator pulses are no longer getting through, the counter stops. The capacitor continues to discharge until it is completely discharged, driving the integrator output to some negative value.

The counter now holds a count that is proportional to the ratio of the reference and input voltages. Since the reference voltage is a fixed value, the count is also directly proportional to the unknown voltage being measured. The count is decoded and displayed in the usual way, and another measurement cycle is begun.

If the unknown voltage is equal to the reference voltage, the counter simply over-ranges a second time. This will usually light a warning LED on the readout panel. Otherwise, the difference between the two counts will produce a numerical readout that is calibrated to indicate the appropriate voltage.

As an example, let's assume we have a reference voltage of one volt and an oscillator frequency of 1 kHz. (1000 Hz). If the unknown voltage is also one volt, the counter, which in our example has a maximum count of 999, will over-range after 1 second (1000 counts). The integrator at this point will have some specific positive voltage, let's say 20 volts. The over-range detector will trigger the input switch to the reference voltage position. Since the reference voltage is the same as the input voltage, it will discharge the capacitor at the same rate—that is, in one second, allowing the counter to cycle through another 1000 counts.

During the complete measurement cycle the counter accepts 2000 input pulses, but its maximum output is 1000 (999 plus the over-range half-digit), so the display reads 1000. The range switch places the decimal point at the appropriate position—1000.

Now, let's change the input voltage being measured to ½ volt. It now takes twice as long to over-range the counter (2 seconds), and the capacitor is charged to only half the level of the earlier example. The output of the integrator stage is 10 volts when the over-range detector triggers S1 to change the input from V_x to V_{ref}.

The reference voltage is still 1 volt, so it discharges the capacitor at the same rate as in the first example. Since the capacitor only has half of its previous charge, it takes the reference voltage half as long as before to reach the 0 volt point, cutting off the comparator/oscillator to counter circuit. Since this time is ½ second, only 500 pulses reach the counter between the time it is over-ranged and and when it is switched off. Therefore, the display reads 0.500 for an input voltage of one half (or 0.5) volt.

Of course, the charging and discharging rates must be extremely linear to insure accuracy in the measurement. However, gradual oscillator frequency drift is not the problem it is with a single-slope converter, since we are concerned with the ratio between the charge and discharge rates, a reasonably stable oscillator will present the same frequency to both halves of the cycle. The precise frequency is not as critical with the dual-slope converter.

Dedicated Digital Voltmeter ICs

A number of dedicated integrated circuits for voltmeter circuits are available. The ICL7106 and ICL7107 are typical. While

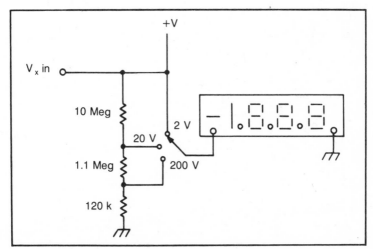

Fig. 18-22. The ICL7106 is a one chip digital voltmeter.

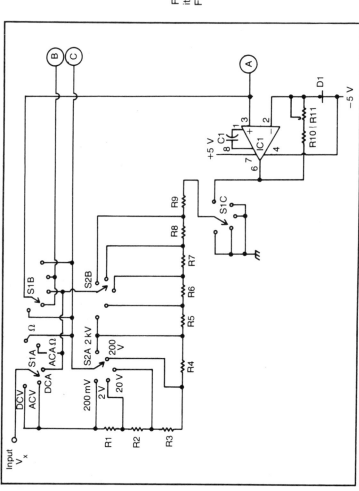

Fig. 18-23. The schematic for a digital multimeter project is shown in Figs. 18-23 through 18-25.

these two devices are functionally similar, they are not inter-changeable. The ICL7106 is designed for use with LED displays, while the ICL7107 is intended for use with LCD readouts. The differing power requirements for the two display types and a few internal differences prevent interchangeability between the two chips.

Besides the display itself, only a range resistor (or range resistors and switch) are needed to construct a functional digital voltmeter. A simplified circuit diagram for such a unit is shown in Fig. 18-22.

The input impedances to these digital meters is quite high preventing problems with loading. Both the ICL7106 and the ICL7107 contain a 3½-digit analog to digital converter, a reference voltage source, a clock oscillator and the appropriate seven-segment decoders and display drivers.

RCA manufactures a pair of ICs that can be combined to make a complete digital voltmeter. They are the CA3161E three-digit analog to digital converter and the CA3162E BCD to seven-segment decoder/display driver. Other dedicated digital voltmeter circuits are also on the market.

A Digital Multimeter Project

The 7107 analog to digital converter, a pair of op amps, and a NOR gate can be used to build a complete digital multimeter that

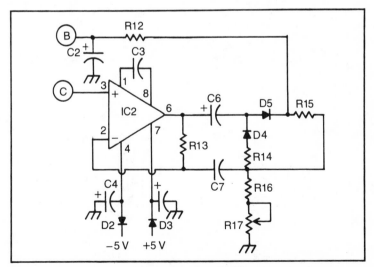

Fig. 18-24. Here is the second portion of the DMM project schematic.

can measure dc and ac voltage and current, and dc resistance. Because of the complexity of the circuit, it is broken down into three parts, and is shown in Figs. 18-23, 18-24, and 18-25. The complete parts list is given in Table 18-2.

Connect matching letters in the three diagrams (A to A, B to B, and so forth). Also be aware that both switch S1 and switch S2 have four connected five-position sections that are shown at separate points throughout the circuit.

In Fig. 18-25, the displays are shown in two places for convenience. In the upper portion of the diagram the connections to the decimal points between the numerals are shown. The seven-segment connections are shown at the bottom. A single 3½-digit display is called for.

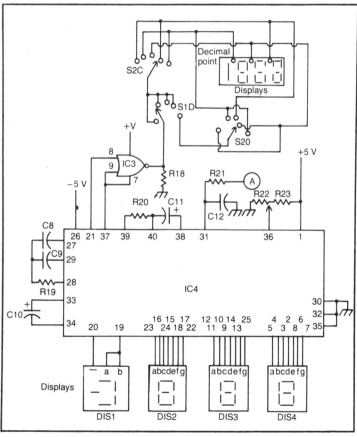

Fig. 18-25. The remaining circuitry for the DMM project.

IC1, IC2	CA3140 op amp, or similar
IC3	7402 quad NOR gate
IC4	7107 A/D converter
R1	9.1 meg resistor *
R2	910 k resistor *
R3	91 k resistor *
R4	9.1 k resistor *
R5	910 ohm resistor *
R6	91 ohm resistor *
R7	9.1 ohm resistor *
R8	0.91 ohm resistor *
R9	0.1 ohm resistor
R10	220 ohm resistor
R11	250 ohm potentiometer (zero set ohms)
R12	51 k resistor
R13	470 k resistor
R14, R15	10 k resistor
R16	2.7 k resistor
R17	5 k potentiometer
R18	2.2 k resistor
R19	51 k resistor
R20	100 k resistor
R21	1 megohm resistor
R22	2.5 k trimpot
R23	22 k resistor
C1	15 pF capacitor
C2,C4,C5,C6	4.7 μF 50 volt electrolytic capacitor
C3	150 pF capacitor
C7	0.1 μF capacitor
C8	0.22 μF capacitor
C9	0.47 μF capacitor
C10	1 μF electrolytic capacitor
C11	100 pF capacitor
C12	0.01 μF capacitor
DIS1	½ digit LED display (with negative sign)
DIS2, DIS3, DIS4	7 segment LED display
D1	CR033 FET regulator
D2, D3, D4	1N4148 diode (or similar)
S1, S2	4 pole, 5 position rotary switch (S1 = function, S2 = range)

* components marked with an asterisk
  should have the lowest possible
  tolerence (1% units are advised).

Switch S1 selects the function to be measured. The five switch positions are labeled *dc volts, ac volts, dc amps, ac amps,* and *ohms.* These positions are shown in the same order in all four sections, even though they are labeled only on S1A. Switch S2 is the range selector. It is labeled for volts (200 mV, 2 V, 20 V, 200 V, and 2 kV), but it is also used for amps and current. Again, the four sections are shown in the same ordered arrangement, although only section S2A is labeled in the diagram.

This project is considerably more complex than any of the other circuits presented so far in this book, but you shouldn't run into any major problems so long as you work slowly and carefully, one section at a time. Double-check all of the wiring before applying power. The use of a printed circuit board and sockets for the ICs is very strongly recommended.

Chapter 19

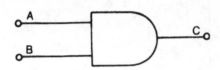

Capacitance Meters

Since the digital voltmeter works by measuring the time it takes to charge (or discharge) a capacitor, it follows that altering the circuitry somewhat we can use a known voltage to find an unknown capacitance.

ANALOG CAPACITOR MEASUREMENT

Capacitors have always been difficult to measure by analog means. A simple go/no-go test may be made with an analog ohmmeter. The capacitor's leads are first shorted together to make sure that it is fully discharged. Then the leads are separated from each other and connected to the ohmmeter's test leads. The needle will shoot down to a very low resistance value as the capacitor accepts current from the meter's reference voltage. Then the pointer will slowly move upscale at a rate determined by the size of the capacitor. When the capacitor is fully charged, the pointer will stop moving, and will indicate the leakage resistance of the capacitor. This should be a fairly high value, but the exact resistance is of little importance. This process is graphed in Fig. 19-1.

Since larger capacitors will charge slower than smaller units (causing the meter pointer to move upscale at a slower rate) this simple test can give you a rough idea of the range of capacitance, but this is extremely imprecise. This test will not work for very small capacitance values, since the charging rate will occur too rapidly for the pointer to have time to respond.

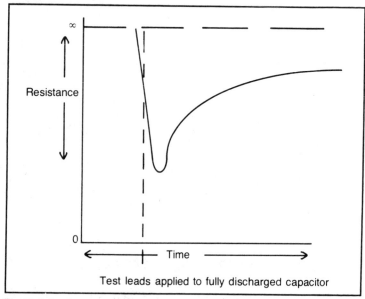

Fig. 19-1. An ohmmeter can be used to give a rough go/no-go test to capacitors.

The main function of this test is to check for shorted capacitors (the resistance reading stays low), and open capacitors (the resistance reading stays very high, or near infinity). Capacitors with excessive leakage will also show a fairly low final resistance level.

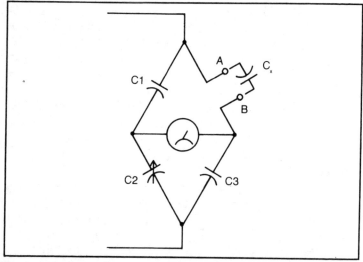

Fig. 19-2. One simple way to measure capacitance is with a bridge circuit.

The ohmmeter test is not suitable for determining the value of an unknown capacitor, or for checking a capacitor for its rated nominal value.

A few analog capacitor meters have appeared from time to time. They generally work on the bridge network principle. A simplified bridge circuit is shown in Fig. 19-2. The unknown capacitor to be measured is connected to points A and B. C1 and C3 have equal values. C2 is a variable capacitor that is adjusted until the meter pointer is precisely centered. This indicates that C2 and C have identical values. A scale is generally mounted on the instruments front panel under C2's control knob. The value can be read from this scale.

This method of measuring capacitors will work, but it is a tedious and time-consuming process at best. The accuracy of the measurements are usually only fair. With modern technology, we can measure capacitances precisely with digital circuitry.

DIGITAL CAPACITANCE METERS

Capacitance values can be readily measured over a wide range using digital ICs. The process is not all that different from the digital voltmeters described in the previous chapter. A basic digital capacitance meter circuit is shown in block diagram form in Fig. 19-3. The first stage is a monostable multivibrator. You should recall that a monostable multivibrator has one stable state. We'll assume the stable state is logic 0.

The output of the monostable multivibrator remains at logic 0 until the circuit is triggered. At that time, the output switches to logic 1 for a specific period of time which is defined by a resistor/

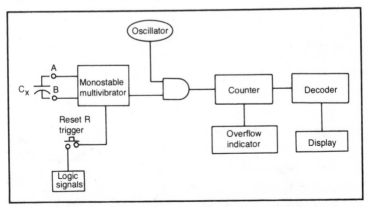

Fig. 19-3. A digital capacitance meter is quite similar to a digital voltmeter.

capacitor combination. After this period of time, the output of the multivibrator returns to its stable state (logic 0).

In this application, the timing resistor is a fixed value. (In some cases, different resistors may be switch-selectable for different ranges.) The timing capacitor is the unknown capacitance being measured. Therefore, the output of the monostable multivibrator goes to logic 1 when triggered for a period of time that is directly proportional to the input capacitance.

The rest of the circuit is very similar to the digital voltmeters described in the last chapter. The output of the monostable multivibrator controls an AND gate which blocks or passes the reference oscillator signal through to the counter stage. The count is checked for over-range, decoded, and displayed.

Since the monostable multivibrator's on time is directly proportional to the unknown input capacitance, the count displayed will also be proportional to the value of C_x. A pushbutton to reset the counter and manually trigger the monostable multivibrator is usually mounted on the front panel of the instrument.

A practical capacitance meter circuit is illustrated in Fig. 19-4. The parts list is given in Table 19-1. Since the voltage across the test points (and therefore through) C_x is less than two volts, the measurement process is safe for virtually any component.

This unit is capable of measuring capacitances from less than 100 pF to well over 1000 μF. If electrolytic capacitors are to be tested, be sure to hook up the meter with the correct polarity. Point A should be attached to the capacitor's positive lead, and point B is negative. Resistor R1 sets the full scale reading for the meter. This component should be a trimpot that is set during calibration, then left alone. Once the necessary resistance is found, the trimpot could be replaced with an appropriate fixed resistor to eliminate the need for periodic recalibration. Smaller resistance values should be used to measure larger capacitances. A 100 k pot could be used for a × 1 scale, while a × 10 scale would be better served with a 5 k potentiometer. Multiple resistances could be set up for switch-selectable ranges.

USES FOR A CAPACITANCE METER

The obvious use for a digital capacitance meter is to determine the value of unmarked capacitors, or to check to make sure a capacitor is true to its marked value. The capacitor's tolerance can be calculated with the following formula:

IC1	555 timer
IC2, IC8	CD4011 quad NAND gate
IC3, IC4, IC6	74C90 decade counter
IC5, IC7	CD4511 BCD to 7-segment decoder
Q1, Q2	NPN transistor
D1	1N4734 diode (or similar)
D2	LED (overflow indicator)
DIS1, DIS2	common cathode seven segment LED display
R1	calibration trimpot—see text
R2, R5	2.7\k ¼ watt resistor
R3, R4, R7	15 k ¼ watt resistor
R6, R8	10 k ¼ watt resistor
R9, R10	1.8 k ¼ watt resistor
R11 - R24	330 ohm ¼ watt resistor
C1, C2	0.047 μF capacitor
C3	0.1 μF capacitor
C4, C5	0.01 μF capacitor
C6	0.0022 μF capacitor
S1	DPDT normally open pushbutton
	—push to clear and test

$$T\% = \frac{ABS(C_M - C_N)}{C_N} \times 100$$

where T is the tolerance (or percent of error) in percentage, C_N is the nominal, or marked value for the capacitor, while C_M is the capacitance value measured.

As an example, let's assume we have a capacitor that is marked 5 μF. When we hook this component up to our digital capacitance meter, we get a reading of 4.357 μF. The tolerance is therefore equal to $(ABX(C_M - C_N)/C_n) \times 100 = (ABS(4.357 - 5)/5) \times 100 = (ABS(-0.643/5) \times 100 = (0.643/5) \times 100 = 0.1286 \times 100 = 12.86\%$. This is not bad for an electrolytic capacitor which often have rather wide tolerances, but it obviously would be a poor choice for any application requiring precision.

A digital capacitance meter can also be put to use in a number of other interesting ways. For instance, this type of instrument can come in very handy for tracking down stray capacitances that can cause problems in many circuits, especially those operating at high frequencies. Printed circuit boards with closely placed traces are often subject to stray capacitance problems.

Components other than capacitors often exhibit internal capacitances that may need to be taken into account when designing precision circuits. For example, in an rf amplifier circuit, the tran-

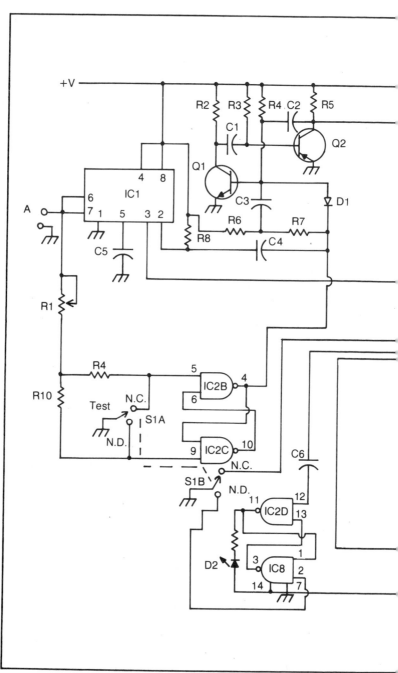

Fig. 19-4. The schematic for a complete digital capacitance meter.

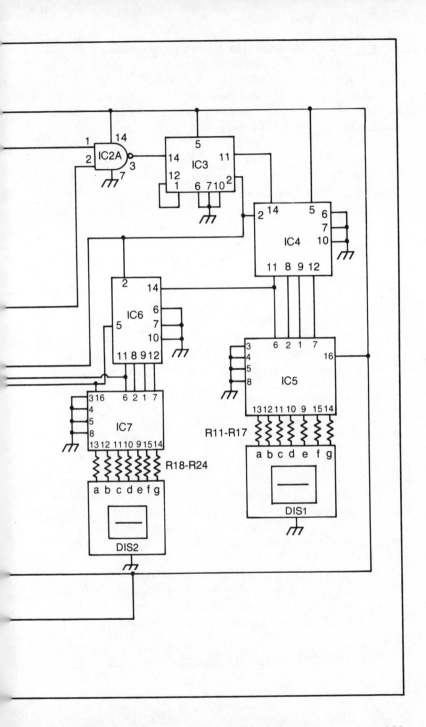

289

sistor base to collector and emitter to collector capacitances may cause instability and/or oscillation in some cases.

Cables that consist of more than a single conductor have a natural capacitance per foot. These cables include coaxial cable, antenna twinlead, and ribbon cables. Since a capacitor is basically two conductors separated by an insulator, multi-line cables naturally behave as long capacitors.

The capacitance per foot of a cable can be determined by measuring a known length of the cable with a capacitance meter, and using the following formula:

$$C_f = \frac{C}{F}$$

where C_f is the cable's capacitance per foot, C is the measured capacitance, and F is the number of feet in the measured sample. C_f and C will always be in the same units. If C is measured in picofarads, C_f will also be in picofarads. Or, if C is in microfarads, C_f will be in microfarads too.

Let's say we have a two and a half foot sample of a cable. When we measure the capacitance of this length, we get a reading of 55 pF. Now, we can easily calculate the capacitance per foot. $C_f = C/F = 55/2.5 = 22$ pF per foot.

By rearranging this formula, we can determine the total capacitance of a length of cable if we know the capacitance per foot, and the length:

$$C_T = F \times C_F$$

where C_T is the total capacitance, F is the length in feet, and C_F is the capacitance per foot. As an example, we'll assume we have a 235-foot length of the cable we worked with in the last example (22 pF per foot). The total capacitance works out to $F \times C_F = 235 \times 22 = 5170$ pF.

Another algebraic manipulation of this same formula allows us to determine how long an unknown length of cable is. This could be necessary if the cable is buried, or embedded in a wall, or otherwise inaccessible for direct measurement. To perform the calculation, we need to know the capacitance per foot, and then take a reading of the cable's total capacitance. The formula is:

$$F = \frac{C_T}{C_F}$$

Let's say we have to find the unknown length of a piece of our 22 pF per foot cable. We measure the total capacitance of the cable to get a reading of 4756 pF. The length of the cable is therefore equal to $C_T/C_F = 4756/22 =$ just over 216 feet and 2 inches.

Another novel application for a digital capacitance meter is as a digital thermometer. For this application a high quality capacitor with a known temperature coefficient is needed. This specification is usually expressed as x parts per million per degree centigrade. For purposes of illustration, we will assume we have a capacitor with a wide temperature coefficient of 100 parts per million per degree centigrade. If this capacitor measures 0.1475 μF at 10 degrees centigrade, at 20 degrees centigrade it should produce a reading of 0.1465 μF. As you can see, a digital capacitance meter can be an extremely handy instrument to have on your workbench.

Chapter 20

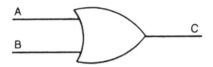

Frequency Meters

In Chapter 18 we learned how a voltage may be measured digitally by counting the number of pulses during the time it takes the voltage to charge a capacitor. Chapter 19 discussed how this technique can be altered to measure an unknown capacitance. There are three elements in each of these measurements. They are a voltage, a capacitance, and a frequency source. Since we have already measured the voltage and the capacitance, it is only reasonable that the same basic techniques can be applied to measuring an unknown frequency. In this chapter we will learn how this is done.

As with capacitance measurement, analog frequency measurement has always been tricky and inconvenient, and usually of poor accuracy. Digital frequency meters, on the other hand, are relatively easy to build, and can be made quite accurate.

DOT-DISPLAY FREQUENCY METER

Before looking at true digital frequency meters, let's take a brief detour to examine a semi-analog circuit. In Chapter 18 we learned about dot and bargraph display voltmeters. In Fig. 20-1 we have a similar circuit that can be used to give a rough reading of frequency. The parts list for this circuit is shown in Table 20-1.

The 555 timer is a reference frequency source. To calibrate the circuit feed is a known (and as accurate as possible) frequency signal, and adjust potentiometer R1 until the LEDs are alternately lit and dark (i.e., on - off - on - off - on - off - etc.).

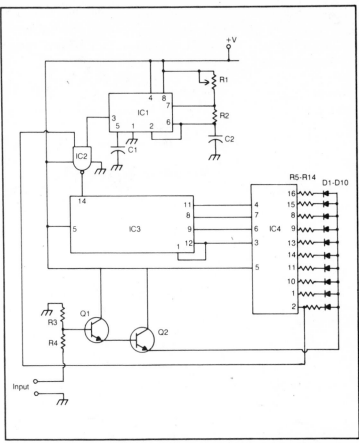

Fig. 20-1. This simple dot display circuit can give a rough indication of frequency.

Table 20-1. Here Is the Parts List for the Dot Display Frequency Meter of Fig. 20-1.

IC1	555 timer
IC2	CD4011 quad NAND gate
IC3	74C90 decade counter
IC4	74C41 BCD/Decimal decoder
Q1, Q2	NPN transistor (2N2222, or equivalent)
D1-D10	LED
R1	100 k potentiometer
R2	1 k resistor
R3	100 k resistor
R4	680 k resistor
R5 - R14	100 ohm resistor
C1	0.01 μF capacitor
C2	0.022 μF capacitor

Now feed in an unknown frequency. Incidentally, for best results, both the reference frequency and the unknown frequency should be rectangle waves. For other signals, a Schmitt trigger input stage may be necessary. The LEDs will light or remain dark in some pattern. Near the center one or more lit LEDs should be grouped together, separated by a dark LED on either side. For example, you may get a pattern like this:

on - off - on - on - off - on - on - on - off - on

Find the largest group of consecutively lit LEDs. In the example, this number is three. Calling this number X, the unknown frequency can be roughly determined with this formula:

$$F_x = \frac{F_R}{X}$$

where F_x is the unknown frequency being measured, F_R is the reference frequency used for calibration, and X is the number of consecutive lit LEDs.

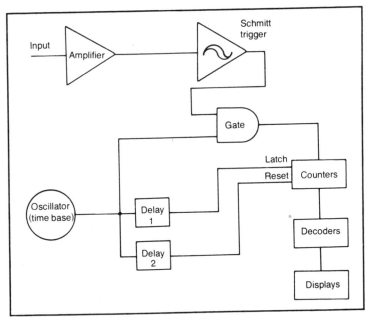

Fig. 20-2. The basic digital frequency meter is another variation on the basic digital voltmeter.

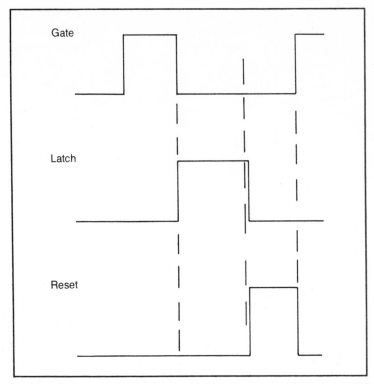

Fig. 20-3. Staggered pulses are required from the timebase oscillator in a digital frequency meter.

Let's assume we used a reference frequency of 750 Hz. If the unknown frequency produces the output pattern in the preceding example, we can conclude that the unknown frequency is approximately $F_R/X = 750/30 = $ roughly 250 Hz.

Obviously, this device is rather complicated to use, as com-

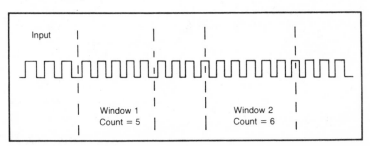

Fig. 20-4. Partial pulses appearing during the "window" counting time may cause some bobbing of the least significant digit.

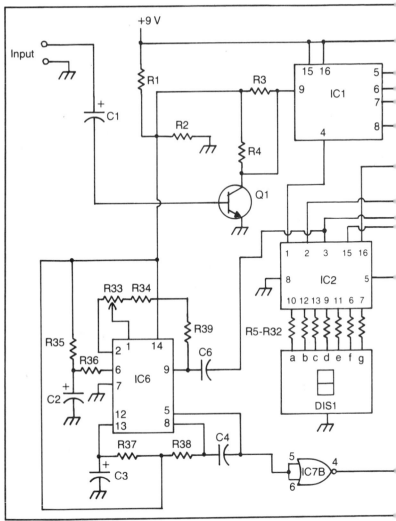

Fig. 20-5. A complete digital frequency meter is shown in this schematic.

pared with devices that readout directly (either analog meters, true dot/bargraphs, or digital displays). Moreover, the resolution is extremely poor. In the example we just looked at the unknown frequency could actually be just about any frequency between about 190 and 375 Hz. Another problem is that the instrument is phase dependent.

The pattern of lit LEDs may fluctuate wildly without providing any meaningful reading at all. Still, even with these drawbacks, this

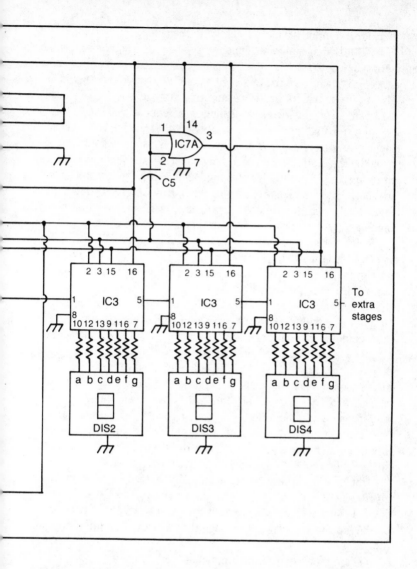

circuit offers an inexpensive way to roughly check frequencies from less than 1 Hz, to about 50 kHz.

HOW DIGITAL FREQUENCY METERS WORK

Most of the commercially available frequency counters around today use the "window" counting method. A sample of the input signal is allowed through a gate. This sample lasts a specific and fixed period of time.. By counting the pulses during this sample period, the input frequency can be determined.

A block diagram for a "window" type digital frequency meter is shown in Fig. 20-2. The input signal is first fed through an amplifier stage to boost the signal to a usable level. An amplifier stage is not always used, but its presence improves the sensitivity of the instrument, allowing lower level signals to be measured accurately.

The next stage of the circuit is a Schmitt trigger to convert any input waveshape to a rectangle wave which can be reliably recognized by the digital circuitry. If only square or rectangle waves are to be measured, the Schmitt trigger stage may be omitted. This processed input signal is fed to one input of an AND gate. The other input to the gating circuit comes from a reference oscillator, or timebase (as it is usually called in this application). The timebase feeds out three signals (or a single signal tapped off with delay circuits, as shown in the diagram). These three timebase signals are synchronized, and their timing relationships are critical. The three signals are illustrated in Fig. 20-3.

The first signal (labeled GATE) is fed to the input of the gating circuit, effectively opening (when logic 1) and closing (when logic 0) the "window," allowing the input pulses to be counted. The second signal, which is delayed until after the first is over, latches the output of the counters, so they can hold their final value long enough to produce a readable display, while the third signal resets the counters to zero for the next measurement cycle. If the output latching was not done, the counter would count up to the appropriate amount, then immediately jump back to 000 and start over, never producing a stationary reading. With the latches, only the desired final count from each measurement cycle is displayed.

Incidentally, the accuracy of most digital frequency counters is given as X% ±1 digit. The least significant digit may bob up and down on successive measurement cycles. This happens because a partial input pulse may get through the "window," as illustrated in Fig. 20-4.

The timebase oscillator must be very precise in its output frequency with as little frequency drift as possible. Crystal oscillators are often used. The input frequency being measured must be higher than the reference frequency. If the input frequency is lower than the reference frequency only 1 or 0 pulses can get through each "window," which obviously would not result in a meaningful reading. To measure lower frequencies, an additional frequency multiplier input stage may be added between the Schmitt trigger and the gate. Similarly, to measure very high frequencies that would over-

Table 20-2. This Is the Parts List for the Digital Frequency Meter Project of Fig. 20-5.

IC1	14583 Schmitt trigger
IC2 - IC5	CD4026 decade counter
IC6	556 dual timer
IC7	CD4011 quad NAND gate
Q1	NPN transistor (2N3302, 2N5826, Motorola HEP-728, Radio Shack RS-2013, or similar)
DIS1 - DIS4	Common cathode seven-segment LED display
R1	22ᵏk resistor
R2	18 k resistor
R3, R39	100 k resistor
R4	10 megohm resistor
R5 - R32	220 ohm resistor
R33	1 megohm trimpot
R34	470 k resistor
R35 - R38	10 k resistor
C1, C3	1 μF 35 volt electrolytic capacitor
C2	10 μF 35 volt electrolytic capacitor
C4, C5, C6	0.001 μF disc capacitor

range the counter stages, a frequency divider stage could be added to drop the input signal to a lower frequency.

Most commercial frequency counters have three to six counter stages for maximum counts of 999 to 999999. Switchable frequency multipliers and/or dividers are also generally included to allow manually selectable ranges. Decimal points may or may not be included in the display readout.

A DIGITAL FREQUENCY METER CIRCUIT

A complete digital frequency meter project is illustrated in Fig. 20-5. Table 20-2 is the parts list for this circuit. Each digit of the display is driven by a CD4026 decade counter (IC2 through IC5). Four digits are indicated in the diagram, but it is a simple matter to extend the display to contain additional digits. Simply connect pin 5 from the last stage to pin 1 of the following stage. The other pins of each IC are connected in the same way as for ICs 2 through 5.

Q1 and IC1 pre-condition the input signal so that it will have an acceptable level and waveshape to be reliably counted by the digital circuits. IC6 is wired as a reference oscillator, whose output frequency can be adjusted with R33, a 1 megohm trimpot. Calibration is done by applying a known frequency source to the input of the circuit and adjusting R33 for the correct reading.

Chapter 21

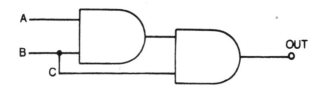

Electronic Thermometers

In Chapter 19 it was mentioned that a digital capacitance meter can be used as an electronic thermometer. In this chapter we will explore electronic thermometers in more depth and learn how they can be made from digital voltmeters and digital frequency meters.

TEMPERATURE SCALES

Before we get down to electronic temperature measurement, we need to be familiar with how temperatures are measured. Three temperature measurement scales are in common use. They are the Fahrenheit, Kelvin, and Celsius (or centigrade) systems. The Fahrenheit scale is the one that is the most familiar to Americans. Except for a few other English speaking countries, the Fahrenheit system is no longer being used through most of the world.

In the Fahrenheit system, water freezes at 32° and boils at 212° at sea level. (Above or below sea level, the changes in air pressure alter the freezing and boiling point temperatures somewhat.)

The reason this system has fallen into general disuse throughout the world is that it is basically a fairly awkward system. 0° Fahrenheit doesn't seem to correspond to anything in particular. The zero point may have originally been determined by the freezing point of a salt water solution, but this is not known, and in any case, 0° Fahrenheit is a rather arbitrary point as far as the day to day world is concerned. Of course, the Fahrenheit system seems natural and fairly simple to us since we have been using it all of our lives. However, familiarity does not necessarily indicate the best system.

Many weather broadcasts today give temperatures both in Fahrenheit and Celsius (or centigrade). The Celsius/centigrade system is the one used throughout most of the modern world. Celsius is the name of the man who invented the system. The term centigrade refers to the fact that this scale has 100 graduations between the freezing point of water, and the boiling point. At sea level, water freezes at 0° centigrade, and boils at 100° centigrade. The scale is based on easily definable, and easy to remember points. Below zero temperatures are below freezing—above zero temperatures are above freezing. What could be simpler?

Many scientific operations use a similar scale, called the Kelvin scale. The graduations are the same as in the Celsius/centigrade system. That is, a change in temperature of 10° Celsius is identical to a change of 10° Kelvin. The difference between the two systems is the 0° point. Instead of the freezing point of water, the Kelvin scale starts at absolute zero (the point at which all molecular action stops). 0° Kelvin equals —459.69° Fahrenheit, or —273.16° Celsius. There are no negative temperatures in the Kelvin scale. This makes complex scientific equations more convenient to work with.

The Celsius scale can easily be converted to the Kelvin scale by just adding 273.16 degrees. This means water freezes at 0 + 273.16, or 273.16° Kelvin. Similarly, water boils at 100° Celsius, or 100 + 273.16, or 373.16° Kelvin.

Converting temperatures between the Celsius and Fahrenheit systems is not quite as straightforward, but it's still not very difficult. For converting from the Celsius system into the Fahrenheit system, this equation can be used:

$$F = (C \times 1.8) + 32$$

where F is the temperature in Fahrenheit, and C is the temperature in Celsius. For example the Fahrenheit equivalent for 100° Celsius is equal to $(100 \times 1.8) + 32 = 180 + 32$, or 212° Fahrenheit. 100° C and 212° F are the boiling point for water, so we know the equation works out right.

For another example, let's convert 53° Celsius into Fahrenheit. Using our equation we find that $(53 \times 1.8) + 32 = 95.4 + 32$, or 127.4 degrees Fahrenheit.

Rearranging the formula allows us to convert temperatures in the other direction—that is, from Fahrenheit to Celsius. The new formula is:

$$C = \frac{(F-32)}{1.8}$$

As an example, 72° Fahrenheit converted into the Celsius scale works out to (72 — 32)/1.8 = 40/1.8, or about 22.22° Celsius.

TEMPERATURE-SENSITIVE COMPONENTS

To build an electronic thermometer we need some kind of sensor to detect changes in temperature. All electronic components are temperature sensitive to some degree (in Chapter 19 we used a capacitor), but some work out better than others. Probably the simplest type of temperature sensor is the thermocouple. A thermocouple is nothing more than a junction of two dissimilar conductors—for instance, a copper wire and a silver wire. When this junction is heated, a voltage proportional to the temperature appears across the wires. This is called the Seebeck effect. The same basic principle is used in many thermostats.

Semiconductors such as diodes and transistors tend to be quite temperature sensitive. These components are often used as temperature transducers. This temperature sensitivity is the primary reason for drift in transistorized amplifiers, oscillators, and other circuits.

When a transistor is used as a temperature sensor, it is gener-

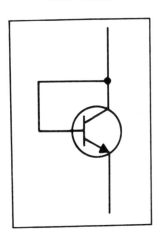

Fig. 21-1. A transistor connected as a diode can be used as a temperature sensor in an electronic thermometer.

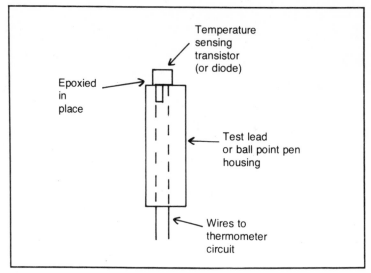

Fig. 21-2. A home-made temperature probe can be mounted in a test lead probe or old ball-point pen housing.

ally connected as a diode, as shown in Fig. 21-1. This is done simply by shorting the collector and base leads. Transistors in metal cases work better as temperature sensors than do those in plastic or epoxy casings. The best transistors for temperature sensing have a very linear base/emitter voltage to collector current function. The sensing works because the base/emitter voltage is determined by the collector current and the temperature of the component. If the collector current is held constant, the base/emitter voltage will vary in direct proportion to the temperature. The voltage can then be easily measured. Because the voltage changes are very small, some kind of amplifier stage will usually be necessary.

If the thermometer is to measure the operating temperature of a circuit, it can be permanently mounted. In most other applications, however, a movable probe will be desirable. The temperature sensing component can be epoxied in the tip of a plastic voltmeter probe, or ball-point pen casing, as shown in Fig. 21-2.

In recent years, specialized temperature dependent components have become available. The most common of these is the thermistor, which is a temperature sensitive resistor. The name comes from *thermal resistor*. Obviously this is a device whose resistance varies in direct proportion to its temperature.

There are two basic types of thermistors. Some thermistors have a positive temperature coefficient. This simply means that as

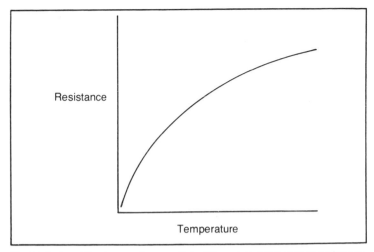

Fig. 21-3. A positive temperature coefficient thermistor increases its resistance as its temperature rises.

the temperature rises, so does the resistance of the component. This is shown in the graph in Fig. 21-3. Other thermistors have a negative temperature coefficient. As the graph in Fig. 21-4 illustrates, these devices work in the opposite direction as their positive coefficient brothers. As the temperature increases, the resistance decreases. Thermistors are generally more useful in analog rather than digital circuits, so we will not dwell on them here. Figure 21-5 shows the schematic symbol for a thermistor.

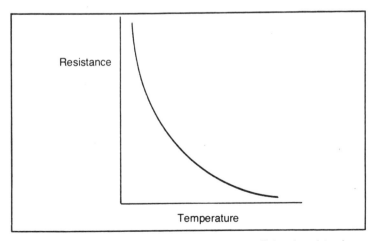

Fig. 21-4. The resistance of a negative temperature coefficient thermistor drops as its temperature is increased.

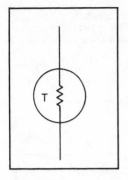

Fig. 21-5. A thermistor is usually represented in schematic diagrams with this symbol.

VOLTMETER THERMOMETERS

Since a diode or transistor can be used to convert a change in temperature to a change in voltage, a simple conversion circuit can be easily combined with a digital voltmeter to create an inexpensive and accurate electronic thermometer. Any of the voltmeter circuits described in Chapter 18 may be used, or you may use a commercially available digital DMM. An analog voltmeter could also be used. Of course the accuracy of your electronic thermometer can be no better than that of your voltmeter.

A simple temperature-to-voltage converter circuit is illustrated in Fig. 21-6. The parts list is given in Table 21-1.

Transistor Q1 is the temperature sensor. It should be a metal

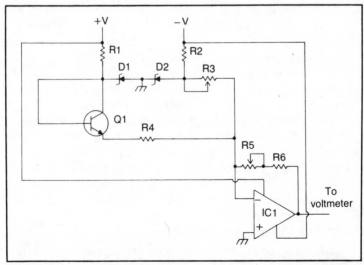

Fig. 21-6. This circuit will convert temperature to a voltage that can be measured with a digital voltmeter.

**Table 21-1. The Temperature-to-Voltage Conversion
Circuit of Fig. 21-6 Is Constructed From These Parts.**

IC1	op amp IC (see text)
Q1	NPN transistor (see text)
D1, D2	identical zener diodes (see text)
R1, R2	10 k resistor
R3	25 k trimpot (see text)
R4	2.2 k resistor (see text)
R5	100 k trimpot (see text)
R6	100 k resistor

case type, and should be mounted in a probe housing as shown in
Fig. 21-2. Almost any NPN transistor may be used. The 2N2222
gives a fairly linear response, and seems to be a pretty good choice.

Zener diodes D1 and D2 set up the reference voltages needed
by the circuit. Their exact value isn't too critical, but it should be
within the 5.1 to 6.8 volt range. Both zener diodes must have the
same voltage rating.

Potentiometer R3 fine tunes the reference voltage to the differential amplifier (IC1). This control should be adjusted for a
reading of 000 when the probe is at 0° centigrade. Potentiometer R5
calibrates the circuit, and should be set for an output of 0.1 volt per
degree centigrade. For example, a temperature of 25.4° centigrade
should produce a reading of 2.54 volts on the voltmeter. Unfortunately, for this simple circuit it is necessary for you to mentally
move the decimal point in the readout one place to the right. For
best results precision potentiometers, such as ten-turn types are
recommended for R3 and R5. Regular trimpots will work, but
precise calibration may prove to be a bit tricky.

IC1 is an op amp to amplify the voltage. Virtually any standard
op amp IC, such as the popular 741 may be used, but a low-noise
type such as the CA3140 will do a much better job.

High quality precision resistors will improve the accuracy of
the circuit. R4, in particular, is fairly critical. A 1% resistor is highly
advisable here. The other fixed resistors (R1, R2, and R6) may be
5% units. Resistors with tolerances of 10% or 20% are not recommended in this project. Tips on calibrating this and other electronic
thermometers will be given later in this chapter.

FREQUENCY METER THERMOMETERS

While a temperature-to-voltage conversion circuit is a natural
for building an electronic thermometer, the voltage changes from
degree to degree are relatively small, leading to inaccuracies

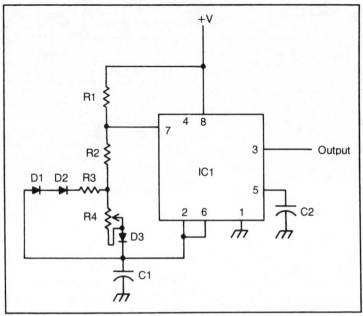

Fig. 21-7. A frequency meter is used to measure the output of this electronic thermometer circuit.

and/or difficulty in displaying the measured value. Often it is a better idea to use the voltage from the temperature sensor to control a VCO (voltage-controlled oscillator) and use a frequency meter to measure the output. Any of the frequency meters discussed in Chapter 20 may be used in this application.

Figure 21-7 shows a very simple circuit for a temperature dependent frequency source. A VCO built around the popular 555 timer IC is the heart of this device. The other required components for this circuit are outlined in Table 21-2.

Diodes D1 and D2 are the temperature sensor. Both should be

**Table 21-2. Here is the Parts List for the
Simple Temperature to Frequency Conversion Circuit of Fig. 21-7.**

IC1	555 timer
D1, D2, D3	1N4148 diode (or similar)
R1	560 ohm resistor
R2	4.7 k resistor
R3	680 ohm resistor
R4	1.5 k trimpot
C1	0.1 μF capacitor
C2	0.01 μF capacitor

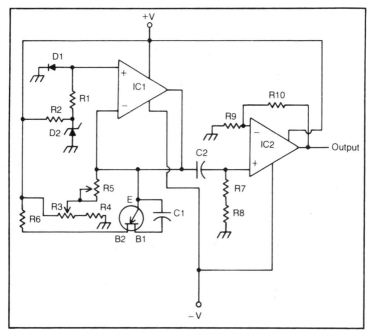

Fig. 21-8. This temperature to frequency converter circuit offers superior performance over the circuit shown in Fig. 21-7.

mounted together at the end of the probe housing. Using two diodes in series like this boosts the voltages generated by the temperature of the sensor diodes.

Potentiometer R4 is used for calibration. As with the voltmeter circuit discussed earlier, low tolerance precision resistors are

Table 21-3. These Components Are Required for the Improved
Temperature to Frequency Conversion Circuit, Which Is Shown in Fig. 21-8.

IC1, IC2	op amp IC (see text)
Q1	UJT (Radio Shack RS-2029, or similar)
D1	1N4148 diode (or similar)
D2	6.2 volt 1 watt zener diode
R1, R4	6.2 k resistor
R2	560 ohm resistor
R3	1 k trimpot
R5	10 k trimpot
R6, R9	1 k resistor
R7	220 k resistor
R8	1 megohm resistor
R10	100 k resistor
C1	0.0047 μF disc capacitor
C2	620 pF capacitor

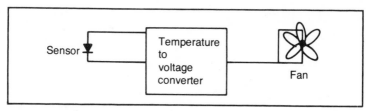

Fig. 21-9. An electronic thermometer can be used to control the voltage to an electric fan.

recommended for the best results and accuracy. You may want to use a high quality capacitor (such as a mylar type) for C1 too. An improved frequency meter/thermometer circuit is shown in Fig. 21-8. Diode D1 is the temperature sensor. The two ICs may be garden variety 741 op amps, but you'll get better results with a higher quality device, such as the CA3140 (see Table 21-3).

If the best components are used, this circuit should have a solid functional range from freezing (0°) to boiling (100°). The temperature is measured in centigrade. The circuit can be adjusted to give readouts that are accurate to the nearest half degree. Potentiometer R3 is used to set the zero point, and R5 is the main calibration control.

CALIBRATING ELECTRONIC THERMOMETERS

Calibrating an electronic thermometer isn't too difficult if you have another good thermometer available. The reference thermometer may be another electronic thermometer circuit, or it may be a traditional mercury in a glass tube type. The first step is to set the zero point, which in the Celsius/centigrade system measured by the circuits in this chapter is the freezing point of water (0° C). To set the zero point, you must first prepare an ice bath. This is nothing more than a container of ice water. Be sure there is plenty of ice. Insert the sensor probe into the water. It's a good idea to put the second thermometer into the water too.

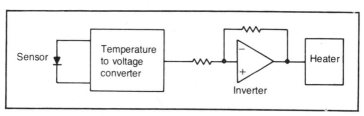

Fig. 21-10. The addition of an inverter stages allows a temperature-to-voltage conversion thermometer to be used in a self-regulating heater.

309

Wait a few minutes until the readings on both thermometers stabilize. Make sure the reference thermometer reads 0° centigrade (or 32° Fahrenheit). Adjust the zero set control (if there is one) on the electronic thermometer you are calibrating. Both thermometers should read 0° centigrade. Let them set in the ice water for about half an hour to make sure the readings are stable. The ice bath should contain enough ice that a significant amount of melting won't take place during this time.

Once you feel the zero point has been correctly set, take the sensor probe and the reference thermometer out of the ice bath and dry them off. Let them sit at room temperature for awhile, until their readings again stabilize. Now calibrate your electronic thermometer for the same reading that is shown on the reference thermometer. An alternative method for the second part of the calibration process is to place the sensor probe in a pan of water on the stove that is just beginning to boil. The water's temperature should be 100° centigrade (or 212° Fahrenheit).

ADDITIONAL USES FOR ELECTRONIC THERMOMETERS

The obvious application for an electronic thermometer circuit is to display the temperature in a meaningful form. But there are several other possible uses. The temperature sensor can be used to control a voltage controlled device. How about letting the thermometer circuit control how much voltage will get through to an electric fan, controlling its speed? This is illustrated in the block diagram of Fig. 21-9.

Similarly, the voltage output of an electronic thermometer

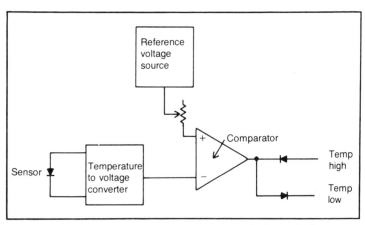

Fig. 21-11. An electronic thermometer can be used as the heart of a thermostat.

circuit can control the temperature of an electric heater. In this case an inverter stage should be added, as shown in Fig. 21-10 so as the temperature of the sensor increases, the voltage being fed through the heater decreases. In other words, we have a sort of self-regulating heat source.

Another obvious application is an electronic thermostat to control switches when the temperature (output voltage) gets higher than and/or lower than a preset temperature. A block diagram of this idea is shown in Fig. 21-11. A manually controlled reference voltage source is used to set the switch-over temperature. Digital gates, SCRs, or relays may be used to switch appropriate external circuits under the control of this circuit.

Chapter 22

Mechanical Counters

In earlier chapters we covered many different digital counter circuits. For the most part, these counted pulses representing electrical characteristics, such as voltage, capacitance, resistance, or frequency. What if we need to electrically monitor some real-world characteristic, such as speed, or the number of objects passing a given point? In this chapter we will explore a few circuits for interfacing digital counter circuits with the outside world.

TOUCH-SWITCH COUNTERS

A touch switch can be a handy real-world input circuit. It is no problem at all to determine the number of times the contacts are touched, simply by counting the number of output pulses. A monostable multivibrator debouncing stage is recommended to minimize the effects of noise and the possibility of the finger shifting position on the touch plate causing multiple pulses.

A basic block diagram for a simple touch-switch counter is shown in Fig. 22-1. A practical circuit is illustrated in Fig. 22-2. The parts list is given in Table 22-1. When you touch your finger across the two touch pads, your body conducts enough electricity to introduce a signal into the circuit which is inverted by IC1A and fed into the monostable mulivibrator (IC2). **It is vitally important not to let any ac power voltage reach the touch pads. Any ac across these pads could cause a painful shock, or even**

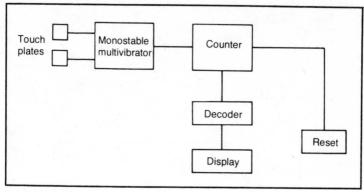

Fig. 22-1. A simple touch switch can be used to trigger a counter.

death. This circuit is recommended for use with battery power only!

The monostable multivibrator acts like a switch debouncer, feeding a clean, reliable signal to the counter (IC3), which is a basic BCD two-digit (00 to 99) decimal counter. The counter can be set back to all zeroes by moving switch S1 momentarily from ground to +V.

The count will continue to increment each time the touch pads are touched, unless the counter is reset via S1, or the maximum count (99) is exceeded.

It's not much trouble to adapt this circuit to include a timebase oscillator, so that the times the touch switch is touched are counted for a fixed period of time. The additional circuitry is shown in Fig. 22-3, and Table 22-2. The original circuit (Fig. 22-2) is broken at

Table 22-1. The Touch Switch Counter Project of Fig. 22-2 Calls for These Parts.

IC1	CD4011 quad NAND gate
IC2	555 timer
IC3	CD4518 dual BCD counter
IC4, IC5	CD4511 BCD to 7-segment decoder/driver
DIS1, DIS2	common cathode seven-segment LED display
	(DIS1 displays the ones, and DIS2 displays the tens)
R1, R2	220 k resistor
R3	3.9 megohm resistor
R4	1 k resistor
R5 - R18	330 ohm resistor
C1, C2	0.01 μF capacitor
S1	SPDT switch

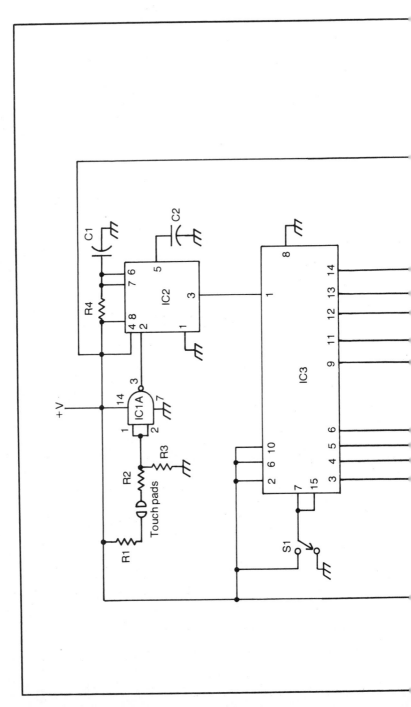

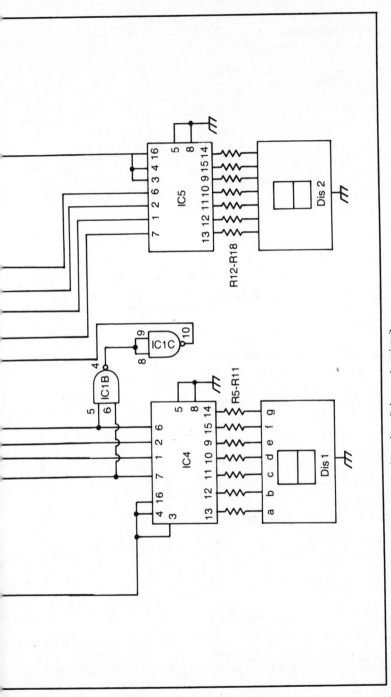

Fig. 22-2. Here is the schematic for a practical touch switch counter circuit.

IC6	555 timer
IC7	CD4011 quad NAND gate
R19	10 k resistor
R20	500 k potentiometer
C3	1 μF 35 volt electrolytic capacitor
C4	0.01 μF capacitor

eliminate S1, and break the circuit at point X (see text)

point X. The signal from IC2 pin 3 is fed to pin 12 of IC1D. This NANDs the input signal with that of the timebase oscillator (IC6). An inverter stage (IC7A) converts the NAND operation into an AND function, so an output signal is fed to the counter input (IC3, pin 1) only when the timebase output is at logic 1.

The timebase oscillator also automatically resets the counter by replacing switch S1 at IC3 pins 7 and 15. The two inverter stages (IC7B and IC7C) are included to slow down the reset signal slightly, allowing time for the count to be displayed before the counter is reset to 00.

MAGNETIC REED-SWITCH COUNTERS

Magnetic reed switches are also good "real-world" inputs for

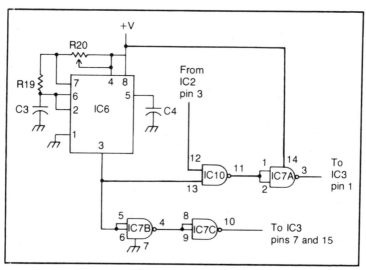

Fig. 22-3. This modification adds a timebase signal to the touch switch counter circuit of Fig. 22-2.

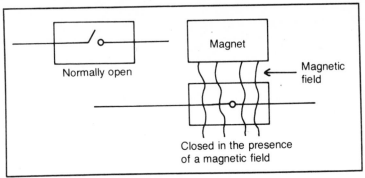

Fig. 22-4. A magnetic reed switch is closed by the presence of a magnetic field.

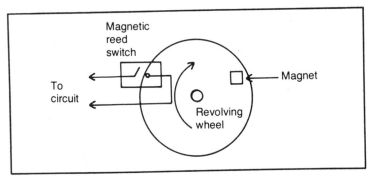

Fig. 22-5. A magnetic reed switch can be used to monitor the motion of a revolving wheel.

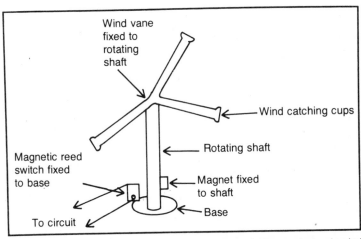

Fig. 22-6. A magnetic switch makes a good starting point for an electronic wind speed vane.

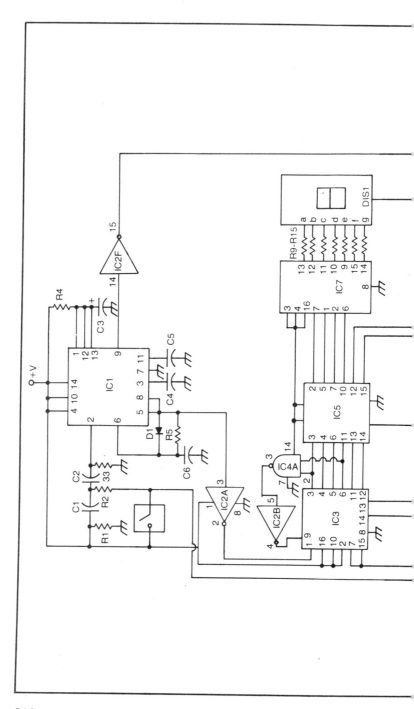

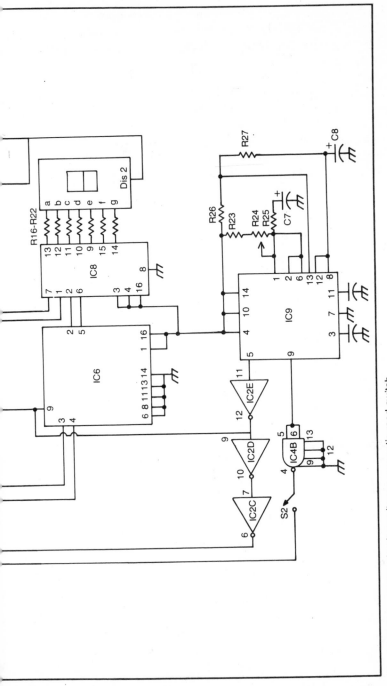

Fig. 22-7. This counter circuit monitors a magnetic reed switch.

digital counters. This device is simply a small enclosed switch that is closed (a few are normally closed rather than normally open) only when a magnetic field is present. The switch is normally opened. See Fig. 22-4. This means we can count how many times a magnet is brought near the switch.

Why would we want to count such a thing? There are actually a great many applications. Let's say we mount a magnet on a revolving wheel, so that it passes a magnetic reed switch once per revolution, as illustrated in Fig. 22-5. We can now keep track of how many times the wheel is revolved. If we compare the number of revolutions with a timebase signal, we can determine how many times the wheel spins around in a given period. In other words, we can measure the speed of the spinning wheel.

This concept can be used to construct an electrode speedometer for a bicycle, go-cart, or even a car. To calibrate such a speedometer circuit, it is necessary to know the circumference (distance around) the wheel. We can calculate how many revolutions there are in a given distance of travel.

Let's try an example. We'll assume we are building a speedometer for a bicycle with 30-inch wheels. This is 2.5 feet. Since there are 5280 feet in a mile, there must be 2112 complete revolutions of the 30-inch wheel per mile. If we update our

**Table 22-3. These Components Are Needed to
Make the Magnetic Reed Switch Counter Project Shown in Fig. 22-7.**

IC1, IC9	556 dual timer
IC2	CD4049 hex inverter
IC3	CD4518 dual BCD counter
IC4	CD4011 quad NAND gate
IC5, IC6	74C174 hex D type flip-flop
IC7, IC8	CD4511 BCD to 7-segment decoder
DIS1, DIS2	Common cathode 7-segment LED display
R1, R3, R4, R5	1 megohm resistor
R2, R8, R25	1 k resistor
R6	47 k resistor
R7	2.2 k resistor
R9 - R22	330 ohm resistor
R23	100 k resistor
R24	500 k potentiometer
R26	56 k resistor
R27	1.5 k resistor
C1, C2, C4, C5, C6	0.01 μF capacitor
C3, C7	4.7 μF 35 volt electrolytic capacitor
C8	2.2 μF 35 volt electrolytic capacitor
S1	SPST magnetic reed switch
S2	SPDT switch

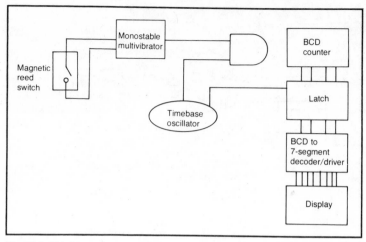

Fig. 22-8. This block diagram simplifies the workings of the magnetic reed switch counter circuit shown in Fig. 22-7.

speedometer reading every 30 seconds, a steady speed of 1 mile per hour, would be 44 feet per 30 second measurement period. Forty-four feet equals 17.6 revolutions per measurement period for a speed of one mile per hour. We can do some rounding off and calibrate out speedometer circuit for a reading of two miles an hour when the wheel revolves 35 times during the measurement period.

A magnetic reed switch counter circuit can also be used to measure wind speed. A wind vane with cups, as shown in Fig. 22-6, can be used to turn a shaft with a magnet mounted on it. The digital circuitry counts the number of shaft revolutions.

Calibrating a wind-speed circuit can be a little tricky. Probably the easiest method is to enlist the aid of a car and a friend to drive. (Do **not** try to drive and perform the calibration by yourself.) A calm day is needed for this calibration procedure. Ideally there should be no wind at all (windspeed of 0 mph). Unfortunately, we can rarely count on ideal conditions. If there is some wind, drive at a 90° angle to the wind direction (or as close to 90° as you can get). Hold the wind vane out the window while the car is moving at a steady rate of speed. Now adjust the circuit's calibration control to get a reading that corresponds to the speed of the car. Of course this method is not exactly precise, but I doubt if the difference between, say, 11.6 mph and 12.4 mph winds is likely to matter much for any consumer application.

A circuit for counting the closures of a magnetic reed switch is illustrated in Fig. 22-7. The parts list is given in Table 22-3. Since

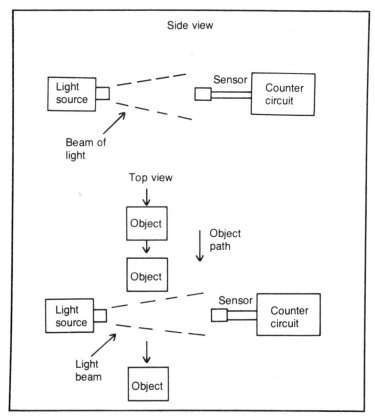

Fig. 22-9. A photoelectric sensor and a light source can be used to count passing objects.

this circuit is fairly complex, it is simplified to block diagram form in Fig. 22-8. Each time the magnetic reed switch is closed, it triggers a switch debouncing monostable multivibrator. This signal is gated with a timebase oscillator and the counting takes place in the usual manner. In the circuit of Fig. 22-7, potentiometer R24 is used to calibrate the readout. Closing S2 puts the counter into a test/reference mode so you can determine when recalibration is required.

PHOTOELECTRIC COUNTERS

Many light sensitive electronic components are available. By mechanically blocking off a light source from a photosensitive device, an object or rotation counter can be easily set up. Figure 22-9 illustrates the basic setup for a photoelectric object counter. A light source and the photosensitive sensor are placed directly opposite each other across the object path. When the object passes this point, the light beam to the sensor is momentarily cutoff, causing it to trigger the counter circuitry. This system can be used to count objects on a conveyor belt, or people passing through a doorway.

A photoelectric rotation counter works on the same principle. A light source and the photosensitive sensor are placed opposite each other. A shaft with a cam is mounted on the revolving wheel, or whatever is being monitored. The shaft rotates with the wheel. Once per revolution, the cam passes between the light source and the sensor, momentarily breaking the light beam. These pulses are then counted by a circuit similar to a frequency meter. This idea is illustrated in Fig. 22-10.

A complete photosensitive counter circuit is shown in Fig. 22-11. The parts list is given in Table 22-4. Transistor Q1 is a phototransistor, and is used as the sensor. Notice that there is no electrical connection to the base of this transistor. The light striking its photosensitive surface provides the base signal.

IC1 is an op amp (such as a 741, or similar device) connected as a comparator. Potentiometer R4 is adjusted so that breaking the beam of light striking the surface of Q1 will cause the comparator to emit an output pulse. This pulse is used to trigger a monostable

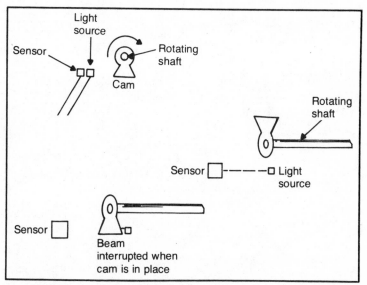

Fig. 22-10. A photoelectric sensor can also be used to monitor revolutions of a moving object.

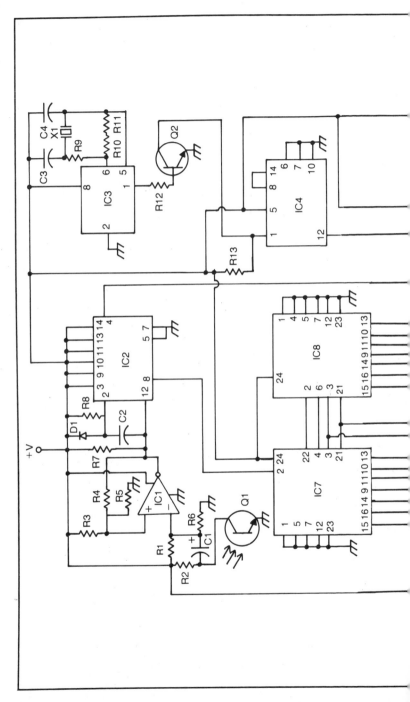

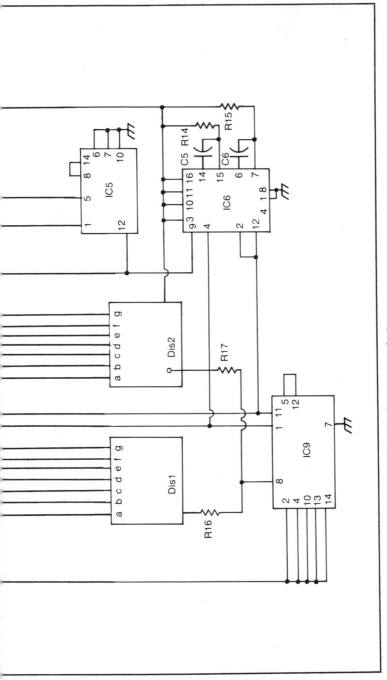

Fig. 22-11. Here is the circuit for a practical photoelectric counter project.

325

IC1	op amp (741 or similar)
IC2	74LS90 J-K flip-flop
IC3	MM5369 60 Hz timebase
IC4, IC5	74LS92 divide by 12 counter
IC6	74LS123 dual monostable multivibrator
IC7, IC8	74LS143 decade counter/decoder/display driver
IC9	74LS74 dual D flip-flop
DIS1, DIS2	common anode 7-segment display with decimal point
D1	1N4148 diode, or similar
Q1	FPT-100 phototransistor
Q2	NPN transistor (almost any type)
R1, R3, R5, R6	220 k resistor
R2	5.6 k resistor
R4	2.5 megohm trimpot
R7, R9	1 k resistor
R8, R14, R15	10 k resistor
R10, R11	10 megohm resistor *
R12	15 k resistor
R13	2.2 k resistor
R16, R17	330 ohm resistor
C1	1 μF capacitor
C2	1000 pF capacitor
C3	30 pF capacitor *
C4	6.2 pF capacitor *
C5, C6	0.033 μF capacitor
X1	3.58 color burst crystal
* see text	

multivibrator (IC2) which cleans up the signal.

IC3 is the timebase oscillator, producing a 60 Hz signal. Since the timebase frequency's stability is more important than its precise frequency, the values of C3, C4, R9, R10, and R11 are not too terribly critical.

The actual counting is performed by IC4 and IC5. IC6 and IC9 control the display's decimal point. Notice that common-anode seven-segment displays are used in this circuit. Common-cathode displays will not work with the 74LS143 decoder/drivers (IC7, and IC8). Transistor Q2 can be almost any "garden-variety" NPN type.

OTHER IDEAS

There are many other ways to let a digital counter monitor physical events. The input sensor could be mercury switches (for position sensing), or a microphone (for counting bursts of sound). Use your imagination. If there is something you need to count, the odds are good that there will be some way to do it with digital electronics.

Chapter 23

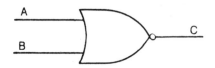

Music-Making Circuits

Circuits that play music or produce sound effects have always been popular among electronics experimenters. Digital circuitry makes many effects easy that would be impractically difficult, or even impossible using just analog techniques.

Any of the circuits presented in this chapter could be used in an electronic music synthesizer, or in toys and games. You won't be building a Moog synthesizer from these projects, but for their cost and simplicity, they can produce countless fascinating and unusual sounds.

TONE GENERATORS

Digital circuits are naturals for generating harmonic-rich rectangle or square waves. The clock oscillators described in earlier chapters are nothing more or less than rectangle/square-wave generators. For musical/sound effects purposes the output frequency must be within the audible range (approximately 40 to 16,000 hertz).

Figure 23-1 shows a simple square-wave oscillator. The potentiometer (R3) can be used to move the output frequency up and down the audible scale. Combining this basic square-wave oscillator with a couple of Schmitt triggers, as illustrated in Fig. 23-2, creates a stepped-wave generator. The parts list for this project is given in Table 23-1.

A stepped wave is made up of several rectangle waves blended

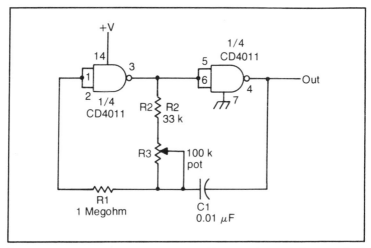

Fig. 23-1. A simple digital square wave oscillator can be made from a pair of NAND gates.

into a sort of staircase shape, as shown in Fig. 23-3. A very bright, brassy sound is created by this waveshape.

Potentiometer R3 controls the number of steps in the output waveform, and thus, the overall tonal quality. The pitch, or frequency, of the output signal is set via potentiometer R4. Nothing is terribly critical in this circuit, so feel free to experiment with different component values. Some very odd effects can be achieved by changing the values of one or both of the capacitors.

Another interesting tone generator circuit is shown in Fig. 23-4. A 555 timer is used to generate a rectangle wave signal, and four D-type flip-flops divide the frequency by half. This gives five simultaneous signals with the following relationship:

Table 23-1. Here Is the Parts List for the Stepped-Wave Generator of Fig. 23-2.

IC1	CD4011 quad NAND gate
IC2	CD4528 dual one shot
IC3	op amp (741, or similar)
R1	1 megohm resistor
R2	47 k resistor
R3, R4, R7	100 k potentiometer
R5, R6	10 k resistor
C1	0.01 μF capacitor *
C2	0.1 μF capacitor *
	* --- see text

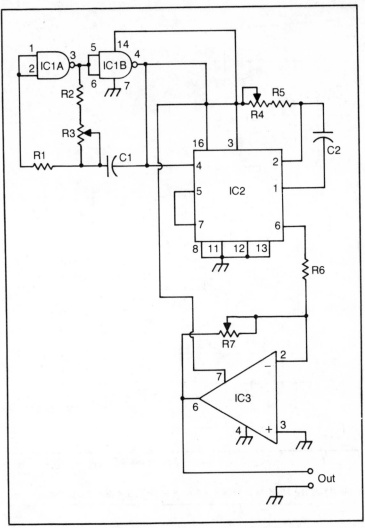

Fig. 23-2. More complex tones can be created with a stepped-wave generator.

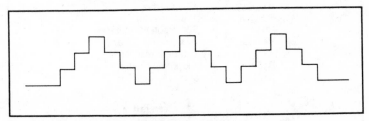

Fig. 23-3. A stepped wave resembles a staircase.

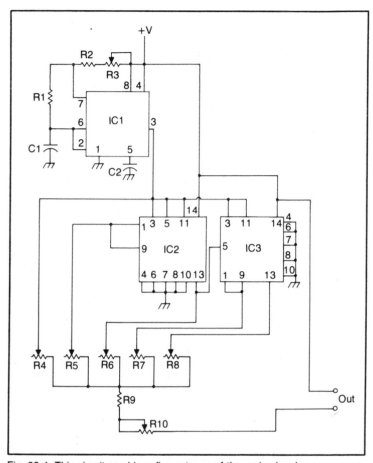

Fig. 23-4. This circuit combines five octaves of the main signal.

Table 23-2. The Multiple Octave Box Circuit of Fig. 23-4 Uses These Components.

IC1	555 timer
IC2, IC3	CD4013 dual D flip-flop
R1	10 k resistor
R2	100 k resistor
R3	500 k potentiometer
R4 - R8	10 k potentiometer
R9	100 ohm resistor *
R10	1 k potentiometer *
C1, C2	0.01 μF capacitor
	* --- see text

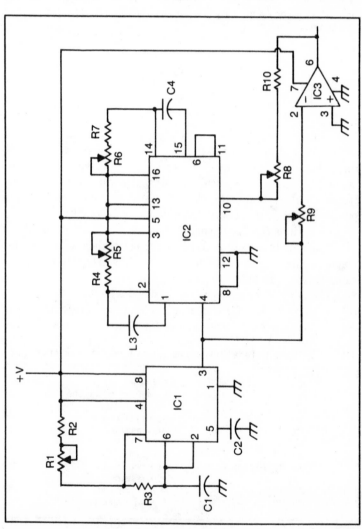

Fig. 23-5. This circuit generates a straight and an out-of-phase signal for unusual effects.

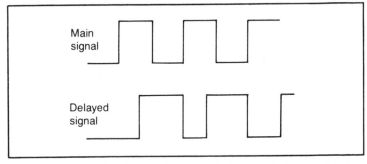

Fig. 23-6. Delaying the main signal in the circuit shown in Fig. 23-5 produces an out-of-phase effect.

$$F$$
$$F/2$$
$$F/4$$
$$F/8$$
$$F/16$$

where F is the frequency of the signal being generated by the 555 timer. Each adjacent frequency is exactly one octave away from its neighbors. The level of each of these signals can be set via potentiometers R4 through R8. A wide variety of combinations are possible, each with a somewhat different tonal quality.

Potentiometer R10 is a master volume control affecting all five of the signals together. If you'd rather not bother with a master volume control, replace R9 and R10 with a single resistor with a value between 120 and 680 ohms.

The pitch of the main frequency signal is set by potentiometer R3. This potentiometer could be replaced by series of switch-selectable resistors to create a rudimentary keyboard, allowing you

Table 23-3. This Is the Parts List for the Dual Phase Generator Circuit of Fig. 23-5.

IC1	555 timer
IC2	CD4528 dual one shot
IC3	op amp (741, or similar)
R1	100 k potentiometer
R2	10 k resistor
R3	3.3 k resistor
R4, R7	1 k resistor
R5, R6, R8, R9	10 k potentiometer
R10	22 k resistor
C1	0.1 μF capacitor
C2, C3, C4	0.01 μF capacitor

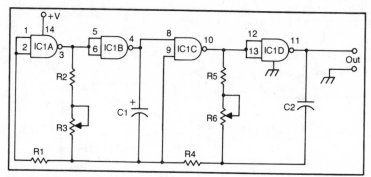

Fig. 23-7. One oscillator can be used to gate a second oscillator on and off.

to play simple tunes on this device. The output of this circuit is a very full and rich tone that is rather suggestive of an organ. The parts list for the project is given in Table 23-2.

The circuit in Fig. 23-5 also splits the main signal for separate manipulation to create new tonal colors. The parts list is given in Table 23-3. IC1 is a 555 timer which generates the main signal. IC2 is a dual monostable multivibrator that delays the signal, producing an interesting out-of-phase effect, as illustrated in Fig. 23-6. Potentiometer R5 determines how long this second signal will be delayed, and R6 sets the width of the delayed pulse.

Finally, both the original signal (through R9) and the delayed signal (through R8) are fed through an op amp mixer/amplifier. By altering the relative levels of the two signals, and trying different settings of R5 and R6, many novel sounds can be generated.

The circuit shown in Fig. 23-7 automatically turns itself on and off at a rate determined by potentiometer R3, and capacitor C1. A tone will be heard at the output in regular pulses. The frequency of

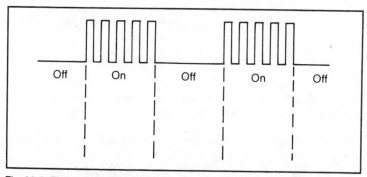

Fig. 23-8. The output of the on/off tone sounder circuit of Fig. 23-7 consists of isolated bursts of an audible tone.

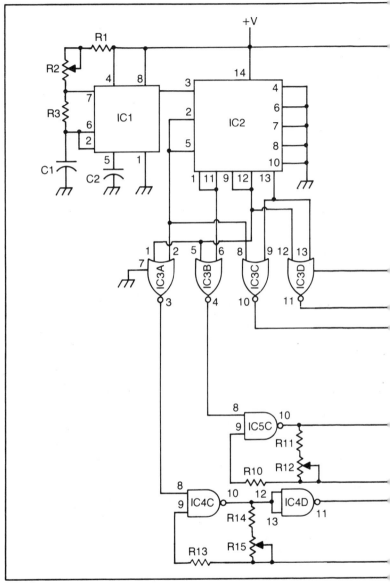

Fig. 23-9. This circuit cycles through four different output tones.

the tone is set with potentiometer R6, and capacitor C2. Figure 23-8
illustrates the way the tone is turned on and off.

All four sections of a CD4011 quad NAND gate IC are used in
this project. Gates A and B are the on/off oscillator, while gates C

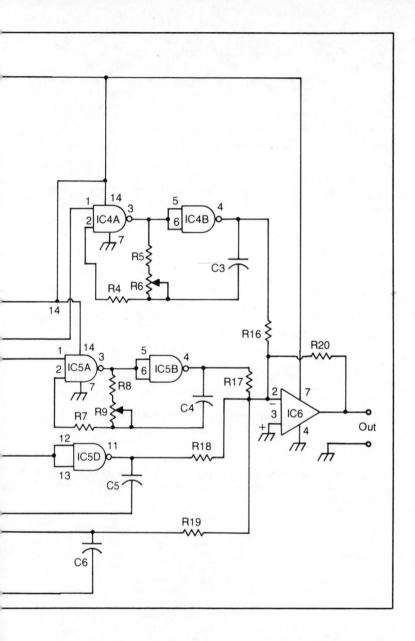

and D form the tone oscillator, which is gated on and off by the first oscillator. Recommended parts values are given in Table 23-4, but nothing is particularly critical. Try substituting different values, especially for the two capacitors.

IC1	CD4011 quad NAND gate
R1	10 megohm resistor
R2	470 k resistor
R3	500 k potentiometer
R4	1 megohm resistor
R5	47 k resistor
R6	100 k potentiometer
C1	0.47 μF capacitor
C2	0.01 μF capacitor

Table 23-4. These Components Are Needed to Build the On/Off Tone Sounder Project Shown in Fig. 23-7.

SEQUENCERS

A sequencer is a kind of electronic music box. It plays through a series of programmed tones automatically. If the playback rate is very fast, the tones will seem to blend into a single complex tone, so sequencers can also be used as complex tone generators.

A basic four tone sequencer circuit is shown in Fig. 23-9. The parts list is given in Table 23-5. The pitch of each of the four tones is set by potentiometers R6, R9, R12, and R15. The 555 timer circuit (IC1) causes the circuit to step repeatedly through the four tones at a rate that can be controlled by potentiometer R2. You can also try using different values for timing capacitor C1.

A more advanced sequencer circuit is illustrated in Fig. 23-10. Table 23-6 is the parts list. Unijunction transistor Q1 and its related circuitry controls the step rate for the circuit at a rate controlled by potentiometer R2 and capacitor C1. For a sequencer, with distinct, separate tonal steps try a capacitor between about 2 and 10 μF. For a complex tone source, a 0.1 μF disc capacitor would be a good choice. There is plenty of room for experimentation.

Table 23-5. The Four Tone Sequencer Project of Fig. 23-9 Calls for These Parts.

IC1	555 timer
IC2	CD4013 dual-D flip-flop
IC3	CD4001 quad NOR gate
IC4, IC5	CD4011 quad NAND gate
IC6	op amp (741, or similar)
R1	2.2 k resistor
R2, R6, R9, R12, R15	100 k potentiometer
R3	33 k resistor
R4, R7, R10, R13	1 megohm resistor
R5, R8, R11, R14	47 k resistor
R16 - R19	10 k resistor
R20	22 k resistor
C1	10 μF 35 volt electrolytic capacitor
C2 - C6	0.01 μF disc capacitor

IC1	7490
IC2	7441
IC3	555
Q1	UJT (2N4891, or similar)
R1	390 k resistor
R2	500 k potentiometer
R3	1 k resistor
R4, R17	100 ohm resistor
R5 - R14, R18	1 k potentiometer
R15	50 k potentiometer
R16	2.2 k resistor
C1	see text
C2	0.22 μF capacitor (see text)

This sequencer has ten steps, each individually set by potentiometers R5 through R14. The voltage passing through each
potentiometer is used to voltage control an oscillator circuit built

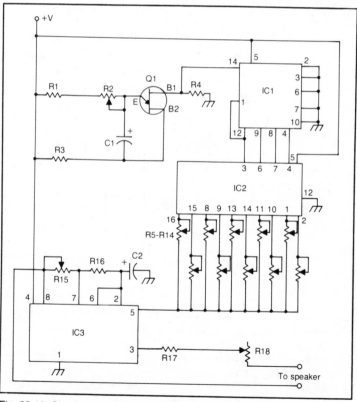

Fig. 23-10. Simple ten note melodies can be programmed on this sequencer
circuit.

337

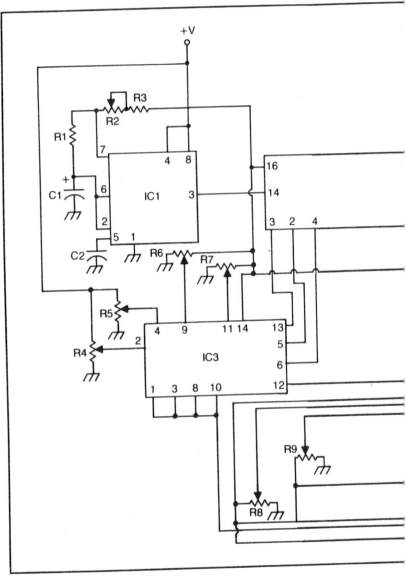

Fig. 23-11. Here is another ten-step sequencer circuit.

around a 555 timer (IC3). Potentiometer R15 is a master pitch control. The pitch range can also be varied by changing capacitor C2. Potentiometer R18 is a master volume control, and may be replaced with a fixed resistor.

Another ten-step sequencer circuit using CMOS ICs is shown

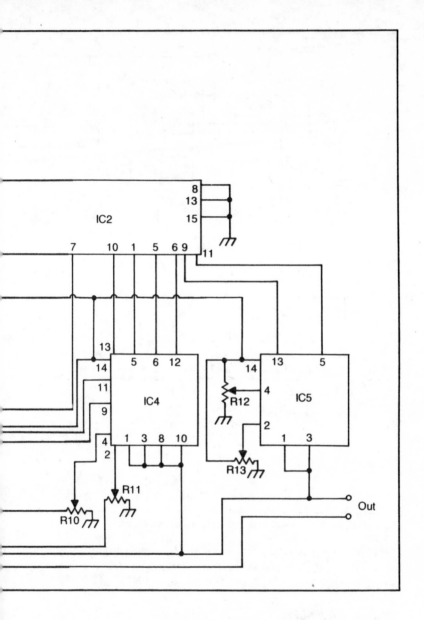

in Fig. 23-11. The parts list is given in Table 23-7. IC1 is a 555 timer whose output is counted by a CD4017 decade counter (IC2). The counter outputs control a series of bilateral switches (IC3 through IC5), controlling the signal through potentiometers R4 through R13. Try experimenting with different values for C1. If you want a

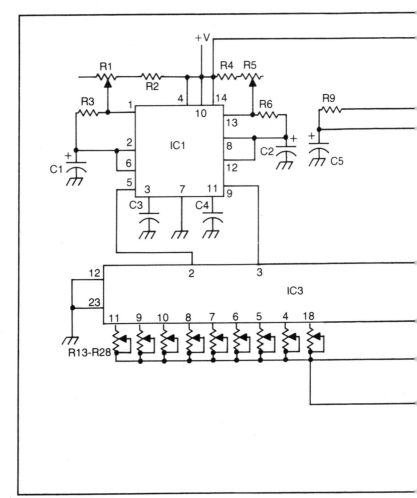

Fig. 23-12. Sixteen tones are randomly played by this circuit.

complex tone generator, use a relatively small capacitor—say, about 0.022 μF. On the other hand, for a sequencer, a larger capacitor, such as a 4.7 μF electrolytic should be used.

RANDOM SEQUENCERS

The sequencer circuits described in the last section will play the same tones over and over again in a continuous pattern unless the control settings are changed. It is also possible to construct circuits that will "make up their own tunes." These circuits will randomly (or pseudo-randomly) select the notes to be played. The

340

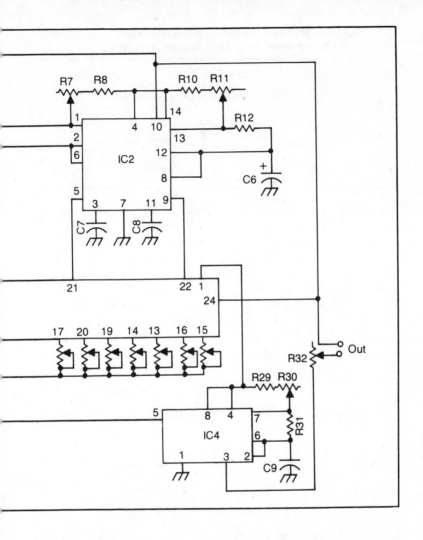

effect is sometimes quite fascinating. Occasionally, a snatch of melody may be heard. Usually, however, the results will sound random.

The circuit of Fig. 23-12 plays 16 tones (selected via potentiometers R13 through R28) in a randomized order. These tones are in the form of voltages to drive the 555 voltage controlled oscillator (IC4). Potentiometer R30 is a master pitch control. R32 sets the output volume.

The heart of this circuit is IC3. This chip is a CD4514 four-to-sixteen demultiplexer. A four-bit binary number fed to the input of

**Table 23-7. The Second 10-Step Sequencer
Project Shown in Fig. 23-11 Requires These Parts.**

IC1	555 timer
IC2	CD4017 decade counter
IC3, IC4, IC5	CD4066 quad bilateral switch
R1	2.2 k resistor
R2	100 k potentiometer
R3	22 k resistor
R4 - R13	5 k potentiometer
C1	see text
C2	0.01 μF disc capacitor

this device causes it to activate one of its 16 outputs, putting a voltage through the appropriate potentiometer.

The four-bit binary number is generated by four independent 555 oscillators. (Actually two 556 dual timer chips are used.) The output of each oscillator serves as one of the four bits in the control number. Since the four oscillators are running at four independent rates (set by R1/C1, R5/C2, R7/C5, or R11/C6) a very irregular pattern of binary numbers will be presented to the inputs of the demultiplexer. See Fig. 23-13.

If tone potentiometers R13 through R28 are set according to a

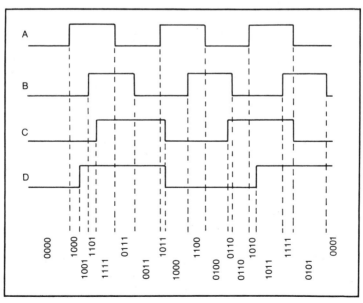

Fig. 23-13. Four square-wave oscillators are used to create a series of random binary numbers.

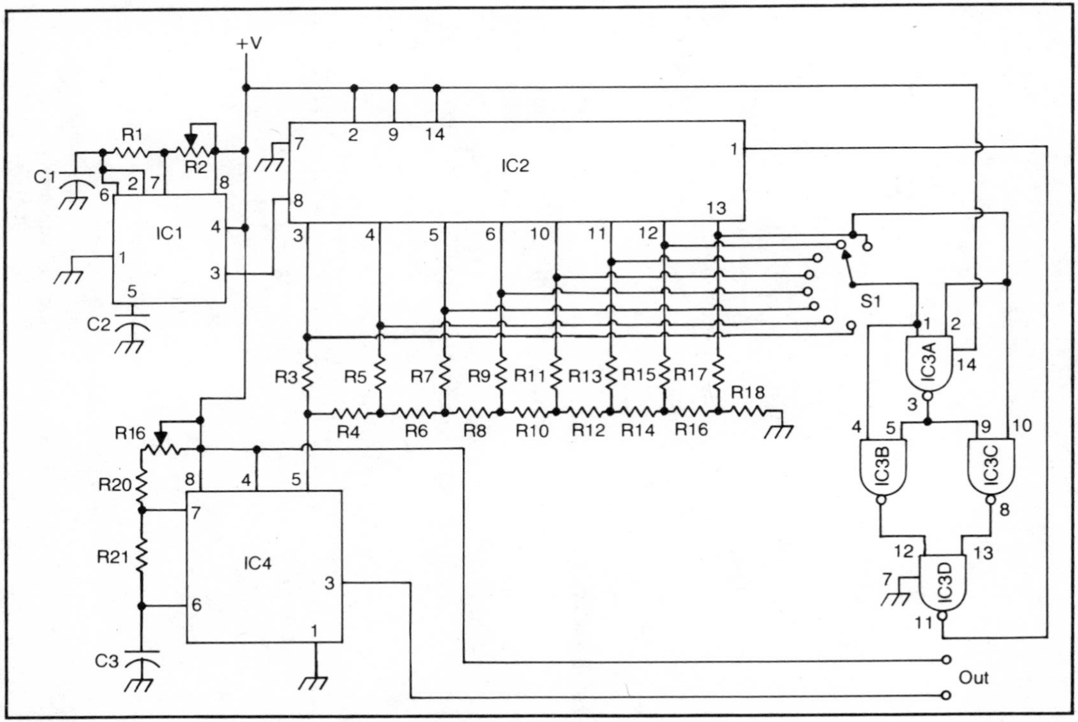

Fig. 23-14. Many random tone sequences can be produced by this circuit.

343

IC1, IC2	556 dual timer
IC3	CD4514 4-to-16 demultiplexer
IC4	55 timer
R1, R5, R7, R11, R13 - R28	10 k potentiometer
R2, R4, R8, R10	22 k resistor
R3, R6, R9, R12, R29	10 k resistor
R30	100 k potentiometer
R31	27 k resistor
R32	5 k potentiometer
C1	33 μF 35 volt electrolytic capacitor
C2, C5	47 μF 35 volt electrolytic capacitor
C3, C4, C7, C8	0.01 μF disc capacitor
C6	100 μF 35 volt electrolytic capacitor
C9	0.1 μF disc capacitor

standard scale, an intriguing pseudo-musical effect can be obtained. The parts list is shown in Table 23-8.

The circuit illustrated in Fig. 23-14, and Table 23-9 will produce an even more random series of tones. The tones heard may or may not correspond to a traditional musical scale. Potentiometer R2 determines the rate at which the tones will change, and R19 sets the over-all pitch range. Changing the position of rotary switch S1 will alter the pattern noticeably.

FREQUENCY DIVIDER

The circuit shown in Fig. 23-15 will accept a high frequency input and divide the frequency by any integer from 1 through 10. The division rate is controlled by rotary switch S1. As an example of how this circuit functions, let's say the input is a 3500 Hz (3.5 kHz) square wave. The output for each position of switch S1 will be:

Table 23-9. This Is the Parts List for the
Random Tone Generator Circuit Shown in Fig. 23-14.

IC1, IC4	555 timer
IC2	74164 8 bit shift register
IC3	7400 quad NAND gate
R1, R20	2.2 k resistor
R2, R19	100 k potentiometer
R3, R5, R7, R9, R11, R13, R15, R17, R18, R21	3.3 k resistor
R4, R6, R8, R10, R12, R14, R16	6.8 k resistor
C1, C3	0.1 μF capacitor
C2	0.01 μF capacitor
S1	1-pole, 8-throw rotary switch

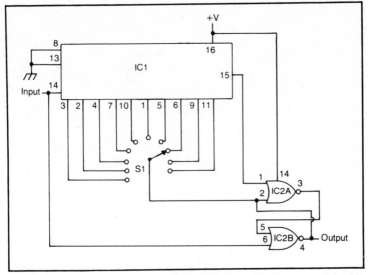

Fig. 23-15. An input frequency can be divided by any integer from 1 to 10 by this circuit.

$$
\begin{aligned}
F/1 &= 3500 \text{ Hz} \\
F/2 &= 1750 \text{ Hz} \\
F/3 &= 1166.7 \text{ Hz} \\
F/4 &= 875 \text{ Hz} \\
F/5 &= 700 \text{ Hz} \\
F/6 &= 583.3 \text{ Hz} \\
F/7 &= 500 \text{ Hz} \\
F/8 &= 437.5 \text{ Hz} \\
F/9 &= 388.9 \text{ Hz} \\
F10/ &= 350 \text{ Hz}
\end{aligned}
$$

Some very interesting effects can be achieved by mixing the original signal with one of the divided values. The parts list for the frequency divider circuit is given in Table 23-10.

FREQUENCY MULTIPLIER

Just to even things out, since we've looked at a frequency

Table 23-10. The Frequency Divider Circuit of Fig. 23-15 Calls for These Components.

IC1	CD4017 decade counter
IC2	CD4001 quad NOR gate
S1	1-pole, 10-throw rotary switch

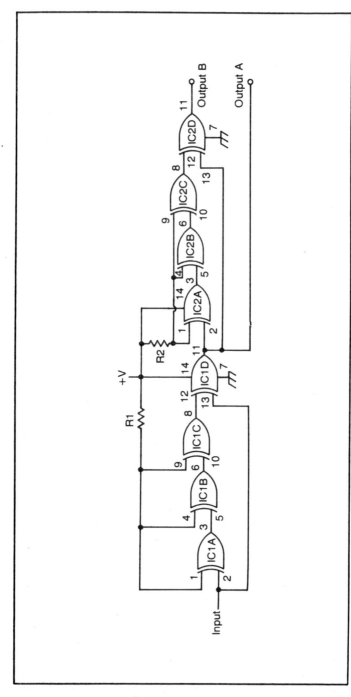

Fig. 23-16. An input frequency can be multiplied by 2 or 4 using this simple circuit.

IC1, IC2	74C86 quad Exclusive-OR gate
R1, R2	2.2 k resistor

divider circuit, Fig. 23-16 is a frequency multiplier circuit. Table 23-11 is the parts list. Output A is twice the frequency of the input, while output B is four times the input frequency. For example, if the input frequency is 1200 Hz, a 2400 Hz signal will appear at output A, and output B will be putting out a 4800 Hz signal. Once again, the most dramatic effects can be achieved by mixing the original, unmultiplied signal with one or both of the outputs of the frequency multiplier circuit.

Chapter 24

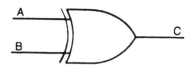

Using Digital Circuits
in Linear Applications

By definition, analog circuits perform linear functions, and digital circuits do not. However, in a number of the projects already shown in this book, analog/linear devices have been used in digital circuits. It is also sometimes possible to force digital devices to imitate linear circuits.

AMPLIFIER

The most basic linear application is amplification. A weak signal at the input appears as a stronger, but otherwise identical signal at the output. A simple linear amplifier can be constructed from three inverter gates, two resistors and a capacitor. The circuit is shown in Fig. 24-1.

The gain of this circuit (i.e., how much larger the output signal will be, compared to the input signal) is determined by the ratio between the two resistors. The gain can be calculated with this simple formula:

$$\text{GAIN} = \frac{R2}{R1}$$

For the best results, both resistors should be relatively large. I wouldn't recommend using anything with a value less than about 100 k. The parts for a linear digital inverter amplifier circuit with a gain of 10 is given in Table 24-1.

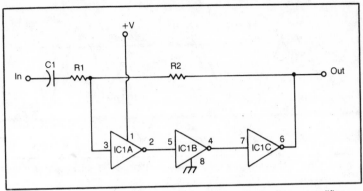

Fig. 24-1. Three digital inverters can be forced to act as a linear amplifier.

Table 24-1. Here Is the Parts List for the Digital Linear Amplifier Circuit of Fig. 24-1.

IC1	CD4049 hex inverter
R1	390 k resistor *
R2	3.9 megohm resistor *
C1	0.01 μF capacitor

* --- see text

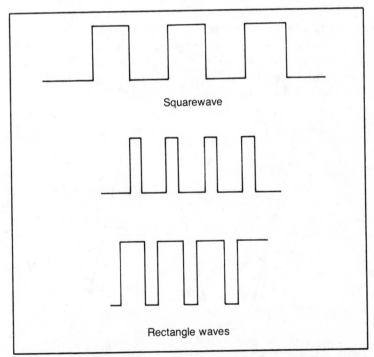

Squarewave

Rectangle waves

Fig. 24-2. Most digital oscillator circuits generate square or rectangle waves.

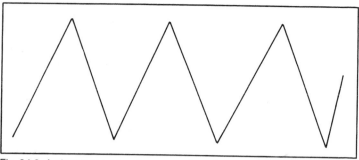

Fig. 24-3. A triangle wave is a linear waveform that can be quite useful in many applications.

OSCILLATOR

We have already seen a number of digital oscillator circuits. But all of these generated rectangle or square waves that simply switch between a high and a low voltage, which is the natural behavior of digital circuits. See Fig. 24-2.

Often other waveshapes may be needed. Figure 24-3 shows a linear waveform known as a triangle, or delta, wave. Another linear waveshape is illustrated in Fig. 24-4. This is a sine wave.

These linear waveforms are difficult, but not impossible to generate by digital means. A reasonable triangle wave can be produced by the circuit shown in Fig. 24-5. Resistor R1 and capacitor C1 are the frequency determining components. The formula for calculating the output frequency is:

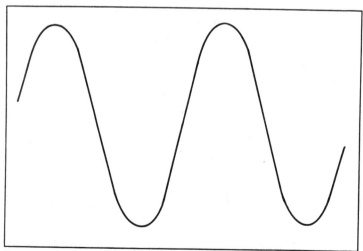

Fig. 24-4. The most basic linear waveform is the harmonic-free sine wave.

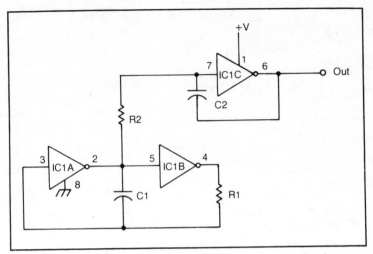

Fig. 24-5. Digital ICs can be used to generate a triangle wave signal.

$$F = \frac{1}{1.4RC}$$

where F is the output frequency, R is the value of R1, and C is the value of C1. Using the component values listed in Table 24-2 will allow the circuit to generate triangle waves at about 1000 Hz.

Sine waves (Fig. 24-4) are usually the most difficult waveforms to generate by either digital or linear means. The ideal sine wave is 100% pure. That is, it consists of just the fundamental frequency with no harmonic content at all. Fortunately, a reasonable approximation will be enough for all but the most critical applications. We can usually settle for a signal that contains a few low-level high harmonics.

The circuit illustrated in Fig. 24-6 generates fair pseudo-sine waves using the phase-shift oscillator technique. Once again, three digital inverter sections are used. All three of the resistors should have identical values, as should the three capacitors. The exact values used will determine the output frequency according to this formula:

Table 24-2. A Digital Triangle Wave Oscillator Can Be Constructed from these Parts (See Fig. 24-5).

IC1	CD4049 hex inverter
R1	15 k resistor *
R2	22 k resistor
C1*, C2	0.05 μF capacitor

* --- see text

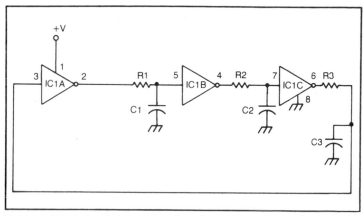

Fig. 24-6. This is the digital version of the linear phase shift oscillator circuit.

$$F = \frac{1}{3.3RC}$$

Table 24-3 lists the parts that should be used for an output frequency of about 1000 Hz.

Table 24-3. The Digital Phase Shift Oscillator of Fig. 24-6 Requires These Components.

IC1	CD4049 hex inverter
R1, R2, R3	3.3 k resistor *
C1, C2, C3	0.1 μF capacitor *
	* --- see text

Even better sine waves can be created from complex stepped waves, as shown in Fig. 24-7. The more steps there are in the wave, the closer it will resemble a true sine wave. This is because the more steps there are, the lower the level, and the higher frequency of the harmonics.

Most of the harmonic content of a stepped wave can be re-

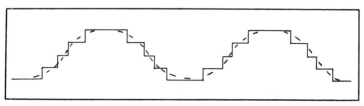

Fig. 24-7. A sine wave can be approximated by filtering a stepped-wave signal.

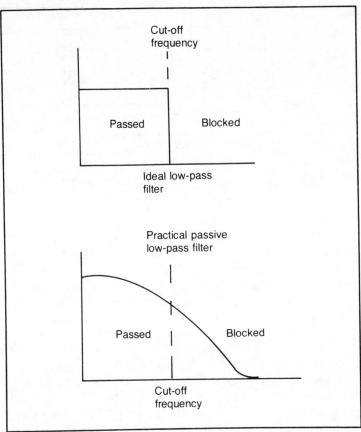

Fig. 24-8. A low-pass filter blocks high frequencies, but allows low frequencies through to the output.

moved from the signal with a low-pass filter. A low-pass filter is a circuit that passes low frequencies, but blocks high frequencies. This is illustrated in Fig. 24-8.

Depending on the component values in the filter circuit, some

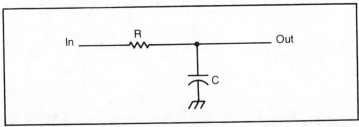

Fig. 24-9. A passive low-pass filter can be made from a resistor and a capacitor.

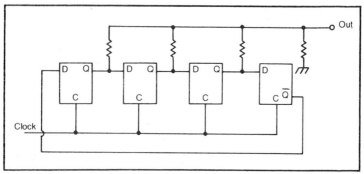

Fig. 24-10. A basic stepped-wave generator can be constructed from cascaded flip-flops.

specific frequency is the cutoff frequency. In an ideal filter all frequencies below the cutoff frequency will be passed completely (with no attenuation), while all frequencies above this point will be completely blocked. This ideal filter response is not possible with real circuitry. Fortunately, in this application, a simple passive filter with a gradual cutoff slope (as shown at the bottom of Fig. 24-8) will do the job reasonably well.

Only two components—a resistor and a capacitor—are needed to create a simple passive low-pass filter. The arrangement of these devices is shown in Fig. 24-9. High frequencies are shunted through the capacitor to ground. Low frequencies, on the other hand, can not pass through a capacitor, so they are fed out through the output. The formula for determining the cutoff frequency for this simple low-pass filter is:

$$F = \frac{159000}{RC}$$

For a 1000 Hz cutoff, you could use a 0.1 μF capacitor and a 15 k resistor.

Table 24-4. Here Is the Parts List for the
Four-Stage Stepped-Wave Generator Circuit Illustrated in Fig. 24-11.

IC1	CD4011 quad NAND gate
IC2	CD4012 quad latch
R1	1.5 megohm resistor
R2, R4, R6	22 k resistor
R3	100 k potentiometer
R5	33 k resistor
R7	2.2 k resistor
C1	0.1 μF capacitor

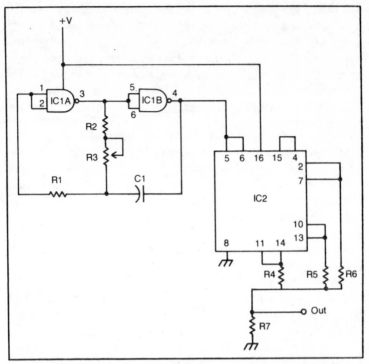

Fig. 24-11. This is the circuit for a four-stage stepped-wave generator.

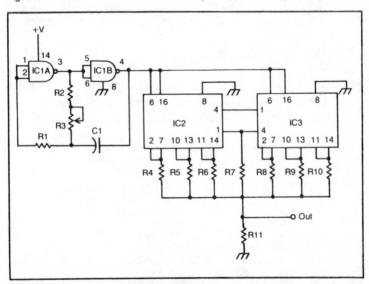

Fig. 24-12. A better sine wave approximation can be obtained by expanding the stepped-wave generator to eight stages.

355

A suitable stepped waveform for digitally synthesizing a sine wave can be made from a string of D-type flip-flops. All of the outputs, except for the last one, are mixed together through a series of resistors. Determining the ideal values for these resistors can be tricky. The values for a four-stage, and an eight-stage circuit will be given shortly. You could also experiment with different resistors. You won't get a sine wave, but you might come up with some interesting waveforms.

The basic stepped-wave generator is illustrated in Fig. 24-10. A practical four-stage generator circuit is shown in Fig. 24-11. The parts list is given in Table 24-4.

IC1 generates a clock frequency which is ten times the output frequency. For instance, if IC1 generates a 10 kHz (10,000 Hz) signal, the output across R7 will be 1 kHz (1000 Hz).

Since more steps in the digital signal improve the similarity to a sine wave, an eight-stage circuit is shown in Fig. 24-12, with the parts list appearing in Table 24-5. This circuit functions in basically the same way as the four-stage circuit of Fig. 24-11, except this time the clock frequency is 16 times the output frequency. A clock frequency of 10 kHz (10,000 Hz) would produce an output frequency of 625 Hz.

DIGITAL FILTERS

Digital circuits can also be brought in to play as filters. Digital filters are often used in communications systems, or in computer analysis of sounds.

Before we turn to an actual digital filter circuit, let's expand the basic passive low-pass filter circuit of Fig. 24-9 into what is known as a commuting filter. In a commuting filter, a number of identical capacitors are switched sequentially in and out of the circuit. A basic commuting filter circuit is shown in Fig. 24-13. The circuit shown

Table 24-5. The Eight-Stage Stepped-Wave
Generator Project of Fig. 24-12 Calls for These Parts.

IC1	CD4011 quad NAND gate
IC2, IC3	CD4042 quad latch
R1	1.5 megohm resistor
R2	47 k resistor
R3	100 k potentiometer
R4, R10	22 k resistor
R5, R9	39 k resistor
R6, R7, R8	56 k resistor
R11	2.2 k resistor

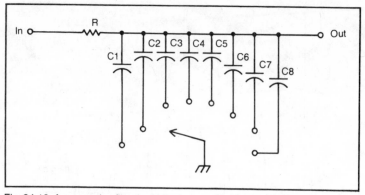

Fig. 24-13. A commuting filter is similar to a passive low-pass filter, with multiple switched capacitors.

here has eight capacitors, so the switch grounds each in turn in this sequence—C1, C2, C3, C4, C5, C6, C7, C8, C1, C2, C3, C4, C5, C6, C7, C8, C1, C2, and so on.

This capacitor switching action changes the operation of the filter from a low-pass to a bandpass type. In a bandpass filter only a specific band, or range of frequencies, is passed. Any frequencies outside of this band (either below or above the pass band) will be blocked. The basic response graph for a bandpass filter is illustrated in Fig. 24-14. The center frequency of the pass band can be found with this formula:

$$F = \frac{1}{2nRC}$$

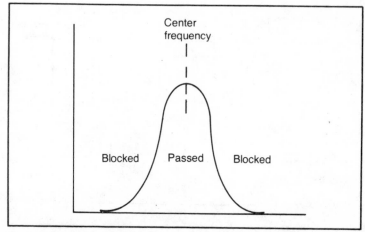

Fig. 24-14. A commuting filter has a band-pass frequency response.

357

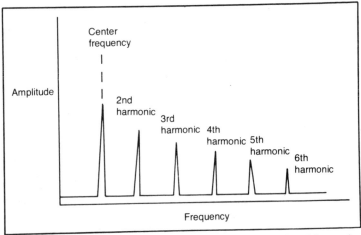

Fig. 24-15. Commuting (or "comb") filters also pass harmonics of the center frequency.

where F is the frequency, R is the resistor value, and C is the capacitance of any one of the capacitors. The number of capacitors, or switch positions, is represented as n.

Actually, there are a number of pass bands. Harmonics (integer

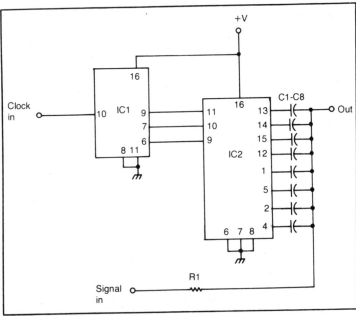

Fig. 24-16. This digital circuit performs a filtering function.

**Table 24-6. These Components Are Needed
to Construct the Digital Filter Circuit of Fig. 24-16.**

IC1	CD4040 BCD ripple counter
IC2	CD4051 BCD to decimal decoder
R1	1 k resistor (experiment)
C1 - C8	0.01 μF capacitor (experiment)

multiples) of the center frequency will also be allowed to pass through the commuting filter. For instance, if the center frequency is 1000 Hz, there will be additional pass bands with center frequencies of 2000 Hz (second harmonic), 3000 Hz (third harmonic), 4000 Hz (fourth harmonic), 5000 Hz (fifth harmonic), and so forth. Each harmonic pass band will be lower in amplitude than the next lower pass band, so the upper harmonics will be eventually filtered out of the output signal. A low-pass filter added to the output of a commuting filter will help get rid of most of these harmonic pass bands if the application requires.

Figure 24-15 shows the frequency response graph for an unmodified commuting filter. Since the multiple pass bands on this graph rather resemble the teeth of a comb, this type of circuit is often referred to as a comb filter. We can use digital circuitry to perform the capacitor switching action, as in the circuit illustrated in Fig. 24-16. Notice that two inputs are required—a clock to drive the counter, and the signal to be filtered. The clock signal should be

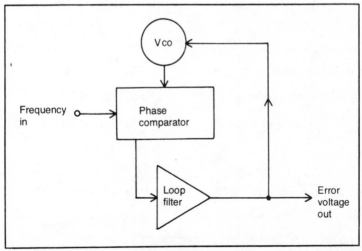

Fig. 24-17. The operating principles for a phase-locked loop are illustrated in this block diagram.

a square or rectangle wave, as with any digital circuit. The signal input to be filtered may be any waveshape. The capacitor values are fairly critical. Components with no more than 10% tolerance rating should be used in this application. Typical component values for this circuit are listed in Table 24-6, but feel free to experiment with different resistor and capacitor values.

A more linear frequency response, and narrower pass bands can be achieved by adding additional decoder stages. The number of stages (capacitors being switched in and out of the circuit) and the clock signal frequency determine the center-frequency pass band, according to this formula:

$$F = \frac{X}{N}$$

where F is the center frequency, X is the clock frequency, and N is the number of capacitor stages.

Almost any input signal may be used for this circuit, but the input signal's peak-to-peak voltage must be less than the voltage powering the digital ICs.

PHASE-LOCKED LOOP

In a number of applications, it is necessary to hold a frequency precisely at a fixed point, and to compensate for any drift that might occur. For example, many FM radios have some kind of auto-tune feature that keeps the receiver from drifting away from the channel

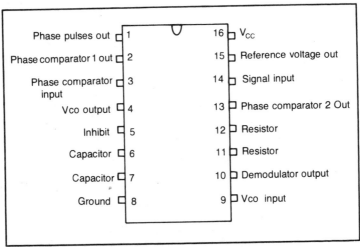

Fig. 24-18. The CD4046 is a digital PLL circuit in IC form.

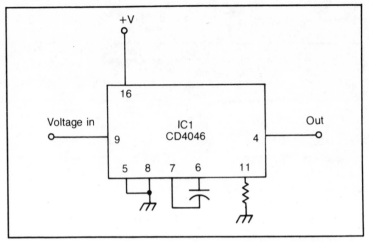

Fig. 24-19. This is the basic CD4046 voltage-controlled oscillator circuit.

(frequency) you're trying to listen to. This process is usually performed by a specialized circuit called a phase-locked loop, or PLL for short.

A phase-locked loop consists primarily of three basic stages, as ilustrated in Fig. 24-17. A voltage-controlled oscillator (or VCO) is tuned to equal the reference frequency. The reference and the VCO signals are checked by a phase-comparator stage. If the two signals are in phase with each other—great. Both are at the same frequency. The output of the phase comparator is 0 volts. But, suppose

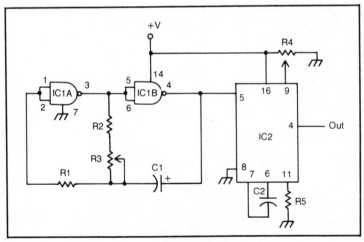

Fig. 24-20. A CD4046 PLL can be used as the heart of a tone-burst circuit.

**Table 24-7. Here is a Typical Parts List
for a CD4046 VCO Circuit, as Shown in Fig. 24-19.**

IC1	CD4046 digital phase locked loop
R1	100 k resistor
C1	0.001 μF capacitor

the two signals don't match. In this case, there will be a voltage at the output of the phase comparator. This signal will be processed through a stage called a loop filter, producing a dc output voltage that is proportional to the amount of error between the two frequencies. This signal is fed back to the control input of the voltage-controlled oscillator, returning it until the error is once again zero.

Ordinarily this is a linear operation taking place in linear circuits. But the CD4046 is a digital phase-locked loop circuit, using CMOS technology. The pinout diagram for this unusual device is shown in Fig. 24-18. It functions in the same basic manner as the linear PLL discussed above, but it can only lock onto input frequencies that are square waves close to the nominal frequency of the VCO when the Exclusive-OR gate phase comparator is used (output pin 2).

But there is a second phase comparator within the CD4046 digital PLL! It can accept an input frequency range of better than 1000:1. However, this subcircuit, which is made up of a complex network of gates and flip-flops, is much more sensitive to noise than is the simpler Exclusive-OR comparator gate. This second phase comparator's output is available at pin 13. Since the two phase comparator stage outputs are separately available, the designer can easily select whichever one will be better suited to the individual application.

The CD4046 digital PLL can be used as a voltage-controlled oscillator with the addition of just a single resistor and a single

Table 24-8. This Is the Parts List for the CD4066 PLL Tone-Burst Circuit of Fig. 24-20.

IC1	CD4011 quad NAND gate
IC2	CD4046 digital phase locked loop
R1	3.3 megohm resistor
R2	47 k resistor
R3	250 k potentiometer
R4	500 k potentiometer
R5	100 k resistor
C1	3.3 μF 35 volt electrolytic capacitor
C2	0.001 μF disc capacitor

capacitor, as shown in Fig. 24-19. Typical parts values are listed in Table 24-7.

Another practical circuit using the CD4046 digital phase-locked loop IC is shown in Fig. 24-20. This circuit will produce bursts of tone, like the on/off tone sounder project introduced in Chapter 23. The parts list for this circuit is given in Table 24-8. Potentiometer R3 will determine the spacing between the tone bursts, while the tone frequency is set by potentiometer R4.

Chapter 25

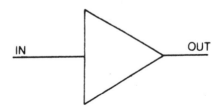

IN   OUT

Problems in Digital Circuits

By now you should realize that digital gates are functionally very simple devices. Each input or output can only have one of two possible states (logic 0 or logic 1), and the relationship between input and output states is clearly defined. Usually. Problems can sneak into digital circuits, and this chapter is intended to give you some tips on finding the causes of and cures for such problems.

Some problems in digital circuits are the same as those found in analog circuits. An incorrect power supply voltage could cause erratic operation, or complete circuit failure. In some cases, some components could be damaged. This is especially true for TTL ICs. Open or shorted filter capacitors are frequent causes of power supply problems. A broken PC board connection or a bad solder joint may cause erratic operation. If the connection is unstable, so that sometimes it is good and sometimes it is bad, the problem may be intermittent, which can be tough to troubleshoot.

If IC sockets are used, you must make sure that all of the pins of each IC are correctly seated. Sometimes one of more pins will get bent under the body of the IC, rather than sliding into the socket. The unconnected pin(s) may be critical for correct operation of the circuit.

All of these problems also occur in analog circuits, and trouble-shooting techniques are the same, so we will not deal with them in depth here. Some potential problems, however, are unique to digital technology, and bear some special discussion.

POWER SUPPLY GLITCHES

A sudden high current draw by a portion of a digital circuit can cause a high current spike, or glitch, to be passed through all of the power supply connections throughout the entire circuit. This could confuse the logic states of the ICs, causing erratic operation, or it may actually damage some of the semiconductor components. This problem is especially prevalent in TTL circuits.

A single TTL gate doesn't draw very much current. It may only sink about 1.6 mA. However, most practical circuits consist of a number of gates, and the current draw can quickly add up. The maximum current is drawn during switching from one logic state to the other. In most gates, there are a pair of transistors. Nominally one will be conducting fully (saturated) and the other will be cut-off completely at all times. During switching from one state to the other, however, there is an instant when both transistors are conducting, greatly increasing the current draw for a brief period. This high current spike appears at all connections to the power supply, since current is equal throughout a series circuit.

Bypassing these narrow high current pulses can be difficult, since ringing can appear across a substantial range of frequencies. The best approach is to include a bypass capacitor across the power supply connections of each and every digital IC in the circuit, as shown in Fig. 25-1. This capacitor should be mounted as close as physically possible to the IC it is protecting. A good value for a bypass capacitor would be from 0.01 μF to about 0.5 μF.

These bypass capacitors are often not shown in schematic diagrams. The projects in this book do not indicate the bypass

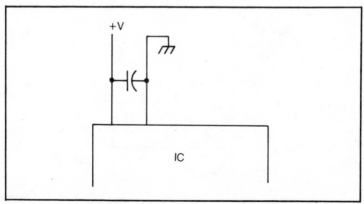

Fig. 25-1. A bypass capacitor should be placed across the power supply bins of a TTL IC.

capacitors in most cases. Sometimes you can get away with not using them, but they are cheap insurance against power supply glitches. The use of bypass capacitors is almost essential in TTL circuits, but its a very good idea with the logic families (such as CMOS) too. A suitable capacitor usually costs about a dime, so they do not significantly affect the over-all cost of the project.

The more gates there are in the circuit, the more important the use of bypass capacitors is. Sometimes you can use a single capacitor to bypass two or three ICs if they are mounted very close to each other, but generally an individual bypass capacitor for each IC in the circuit is advisable.

PROPAGATION DELAY

Another problem that frequently plagues designers of digital electronics circuits is propagation delay. We all too often tend to forget that it takes a finite amount of time for a digital gate's output

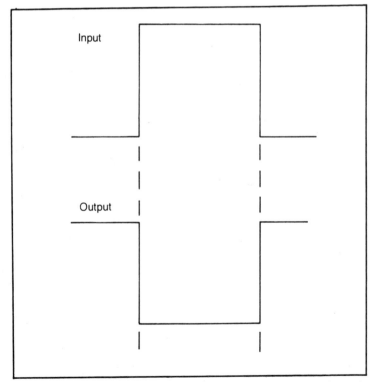

Fig. 25-2. An ideal inverter's output changes instantly in response to a change in the input signal.

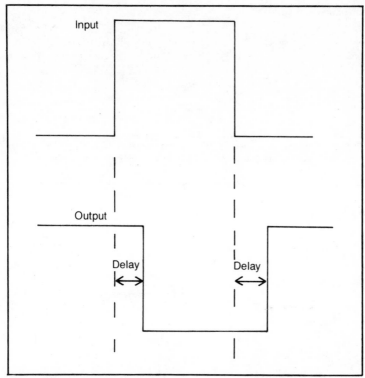

Fig. 25-3. In a practical inverter, there is a brief delay before the output can respond to a change in the input signal.

to respond to a change in the input signal. Figure 25-2 shows the idealized input and output signals for a digital inverter. We show the output changing simultaneously with the input. For many applications, we can accept this idealized image. However, in practical gates, there is a brief delay before the output responds, as shown in Fig. 25-3. This delay is a tiny fraction of a second, but it does occur, and in some circuits it can make all the difference in the world.

Consider the hypothetical digital gating circuit illustrated in Fig. 25-4. Obviously, this is not a practical circuit. It is for purposes

Table 25-1. Ideally the Demonstration
Circuit of Fig. 25-1 Should Exhibit this Truth Table.

Input	Inverter Outputs A B C D	AND Gate Output
0	1 0 1 0	0
1	0 1 0 1	1

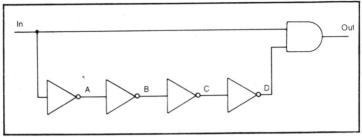

Fig. 25-4. This circuit is used to demonstrate the potential problems caused by propagation delay, as explained in the text.

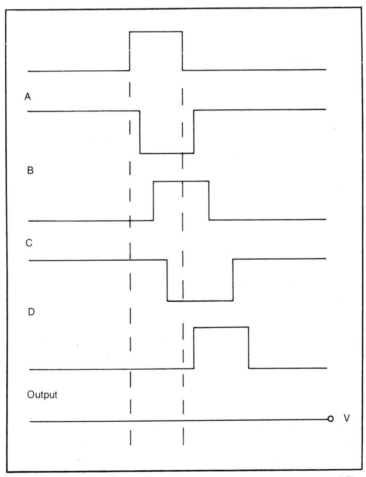

Fig. 25-5. Here are typical timing signals for the demonstration circuit of Fig. 25-4.

of demonstration. The input signal is split into two paths. One portion of the input signal goes directly to an AND gate. The other portion passes through four inverters before being applied to the AND gate's other input.

Since each pair of inverters logically cancel each other out, the signals appearing at the inputs of the AND gate should be identical. The output of the entire circuit should be the same as the input, as indicated in the truth table shown in Table 25-1.

But what happens if the input signal is a brief pulse? Figure 25-5 illustrates the logic signals appearing throughout the circuit with a brief pulse input. The propagation delay of each device is taken into account.

While the propagation delay of a single inverter may appear inconsequential, notice the cumulative effect. The total propagation time of the four inverters together is longer than the input pulse. By the time the signal from the longer, inverter path reaches the input of the AND gate, the original signal is gone. The output remains at logic 0. The input pulse is never indicated by the circuit output at all.

Propagation delay problems can sometimes be very tricky to locate. As in this example, you may find that the circuit as a whole does not operate properly, even though each individual stage is doing its job. Clearly the designer should take the propagation delay of the circuit stages into account when creating a circuit. If he doesn't, the timing errors in all probability will come back to haunt him sooner or later.

If you build a digital circuit that starts behaving in an oddball way, suspect propagation delay as a likely culprit. Typical propagation delay figures are usually indicated on manufacturer's data sheets for individual ICs. Unfortunately, for technical reasons, it is impossible for the manufacturer to define the precise propagation delay. Otherwise identical devices (from different batches) could very well exhibit different amounts of propagation delay. To further complicate matters, the gate switching time can be affected by external factors, such as temperature, and load capacitance. Generally, IC data sheets list minimum and maximum propagation delay times for a given type device. You can reasonably expect the actual value to be somewhere within this stated range.

When designing, assume the worst possible case. If long propagation delays will cause you the most circuit operation problems, assume that all devices will exhibit the maximum propagation delay time, and compensate for this value.

In some circuits, a short propagation delay may cause more problems than a long one. In such a case, assume that all devices will be operating with their minimum stated propagation delay (the worst case for this circuit). Some designers take advantage of propagation delays in their circuit designs. If a portion of split signal needs to be delayed briefly for some reason, they try to let the chips' propagation delay's do the job for them. This is a risky business at best, and probably should be avoided whenever possible.

Occasionally (as with a couple of the circuits presented in earlier chapters of this book), using propagation delay for circuit operation may turn out to be the practical way to go. In a couple of the counter circuits, I used the propagation delay of a string of inverters to hold back the reset signal until the counter circuit has had a chance to lock onto its final value. Since only a very brief instant of delay is required (although a longer delay wouldn't hurt) we can get away with making propagation delay work for us, since other methods of delaying the reset signal could raise the over-all circuit cost. The back-to-back inverter stages that cancel each other out and don't affect the logical operation of the circuit are used as digital delay lines. In the design, we make sure that there will be more delay than we absolutely need, even under worst-case conditions.

Usually, however, longer propagation delays will be more trouble than short ones. This is especially true when a split signal flows through paths containing a drastically unequal number of gates. One way to compensate for this would be to lengthen the shorter path by adding "logically neutral" gates, like buffers, or back-to-back inverters. But be warned—there can still be problems! Even if both paths have exactly the same number of gates, but the gates in path A happen to all have the minimum propagation delay for their type device, and the gates in path B happen to all have maximum propagation delay. The total difference (especially for long paths) would have a cumulatively disasterous effect.

Never assume that the total errors in a circuit will cancel each other out. A good circuit designer is a confirmed pessimist. Always assume the worst possible operation conditions and the worst possible specifications for your specific application. It's far better to over-compensate for potential errors than to under-compensate for existing errors. OR and NOR gates tend to be less susceptible to errors stemming from propagation delay than AND or NAND gates.

Synchronized circuits (i.e., the same clock signal being applied to all the individual stages) will be less affected by propagation

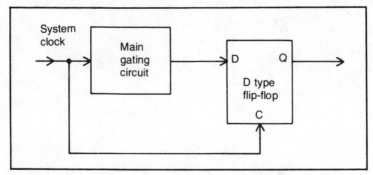

Fig. 25-6. A flip-flop can be used to help compensate for propagation delay.

delay errors than will non-synchronous circuits, such as ripple counters. In some applications, taking advantage of the enable inputs on many MSI and LSI ICs will cut down on "glitches."

Remember that some logic families operate slower than others. For example, CMOS gates tend to have longer propagation times than similar TTL gates. This can be of importance in circuits combining different logic families, as discussed in Chapter 12.

In a pinch, you can add a delay compensation circuit, such as the flip-flop circuit shown in Fig. 25-6, or the RC/inverter circuits illustrated in Fig. 25-7. Note that the circuit in Fig. 25-7A, the signal will be inverted, but in Fig. 25-7B the inversions cancel each other out, so that the output mimics the input signal. The RC/inverter trick is best restricted to CMOS circuits. In TTL circuits it may cause impedance problems.

NOISE

Noise is any unwanted signal. Obviously this can cause problems if a significant noise signal gets into your circuit, and the circuit is unable to distinguish the noise from the desired signal. Bypass capacitors and Schmitt triggers (see Chapter 6) can help cut down significantly on noise problems in digital circuits. A closely related problem is mechanical switch bouncing. A switch debouncer circuit was presented and discussed in Chapter 15.

TROUBLESHOOTING

So far in this chapter we have been concentrating on problems of concern to the circuit designer or those that show up when the circuit is constructed. We all know, however, that sooner or later good circuits can go bad and require servicing. Troubleshooting

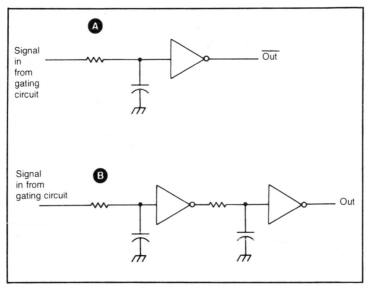

Fig. 25-7. This RC/inverter circuit can be used to minimize propagation delay errors.

digital electronics circuitry is made far easier with a few inexpensive special-purpose test instruments.

Logic Probe

A logic probe is probably the single most useful piece of specialized digital test equipment. A logic probe project that will do for most troubleshooting of the circuits presented in this book was included in Chapter 15.

A number of commercially available logic probes have appeared on the market over the years. Some of these are fairly simple and inexpensive devices, along the lines of the project discussed in Chapter 15. Other commercial logic probes include a number of specialized features, such as indicators for repeating pulses, and erroneous voltages between the defined logic levels.

Many digital circuits work with very rapid pulses. The indicator LED may blink on and off too fast for the eye to catch. For this reason, several deluxe logic probes include "pulse stretchers." A pulse stretcher is nothing more than a monostable multivibrator. The quick pulse from the circuit triggers the monostable multivibrator, which is set up to produce a longer pulse at its output, as shown in Fig. 25-8. A logic probe can be used to find the current logic state of any single point within a digital electronics circuit.

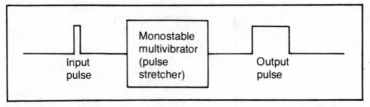

Fig. 25-8. Brief pulses can be "stretched" to overcome problems stemming from propagation delay.

Logic Clips

Ordinary logic probes allow you to check one IC pin at a time. As useful as this can be, sometimes you may need to simultaneously monitor both the inputs and the outputs of a digital IC. The concept of the logic probe can be expanded to come up with a device known as a logic clip. A logic clip is essentially a multiple logic probe. It is connected to the IC with a special clamp that vaguely resembles a clothes pin. This clip, called a Glomper clip, makes simultaneous contact with all of the pins on a single IC chip.

Usually a 16-pin Glomper clip is used for the greatest versatility. If an 8-pin or 14-pin chip is to be monitored, the extra connectors are allowed to just hang unconnected over the end of the shorter IC. Glomper clips can only be used with ICs in DIP packages.

A regular logic probe can fit into many places and get to connections where a Glomper clip won't fit. Moreover, very rapidly

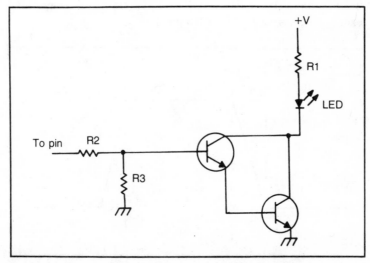

Fig. 25-9. This is a typical one-pin circuit for use in a logic clip.

373

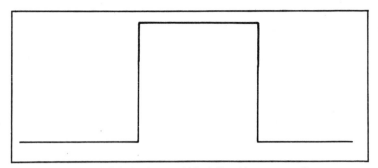

Fig. 25-10. An ideal digital pulse would look like this on an oscilloscope.

changing logic states could be rather confusing on a logic clip. With a standard logic probe you can concentrate on one logic signal at a time. However, the logic clip is great for monitoring an entire IC, especially when the logic states are static (unchanging), or changing at a relatively low rate.

Most logic clips have only a single LED indicator for each pin. This means that there is a solid indication for only one of the two possible logic states. For example, if the LED lights on a logic 1 signal, a logic 0, a ground connection, or no connection at all would all leave the LED dark, creating some ambiguity. Yet, the one LED per pin keeps the cost and instrument size down, and it will give reasonably reliable readings in most cases.

Each pin drives its own logic probe circuit. A typical circuit for use in a logic clip is shown in Fig. 25-9. All of the components, except R1 are repeated for each pin connection.

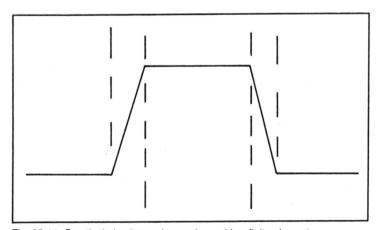

Fig. 25-11. Practical circuits produce pulses with a finite slew rate.

374

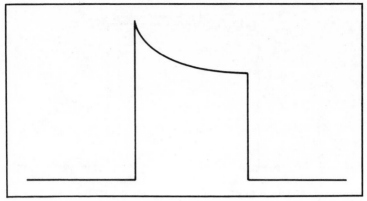

Fig. 25-12. This pulse suffers from over-shoot.

Logic Pulser

In analog servicing, a signal injector is often used to help find the defective circuit stage. A signal injector is an instrument that can feed a known signal into any desired point in the circuit under test. Since the signal's characteristics are known and are under the control of the technician, tracing out the action of the various circuit stages is made much easier than might otherwise be possible. This same technique can also be useful in servicing digital electronics circuits although an analog signal tracer would not be suitable.

The digital equivalent to the signal injector is called a logic pulser. As the name implies, this is a circuit that emits correct logic pulses. A square wave oscillator can generate a string of pulses at the desired frequency. In some cases, it will be more useful to

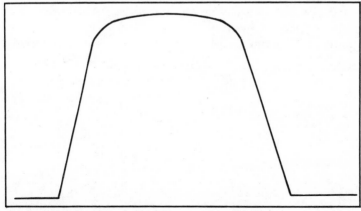

Fig. 25-13. Under-shoot can cause a pulse to look like this.

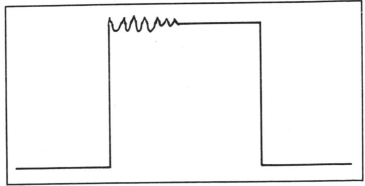

Fig. 25-14. Fast switching times can cause oscillations, or "RINGING" to ride on the pulse signals.

monitor the results of a single pulse. To accomplish this, most logic pulsers also include a manual switch which can trigger a monostable multivibrator, producing a single pulse at the desired logic levels, and with the desired duration. Many logic pulsers include built-in logic probes to monitor their signals.

Other Test Equipment

Large laboratories will have more advanced test equipment, such as logic analyzers which can compare various signals throughout the circuit and combine them into a meaningful readout. Such devices are beyond the point of affordability for most hobbyists, and even most service technicians. Fortunately, such deluxe equipment is really something of a luxury that would make some jobs easier, but is not absolutely essential.

Some analog test equipment can be used when working on digital equipment. An analog VOM or VTVM can monitor static, or slowly changing logic signals. If the pointer indicates a voltage near

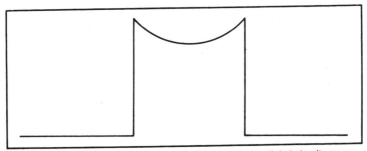

Fig. 25-15. A distorted pulse may not be recognized by digital circuitry.

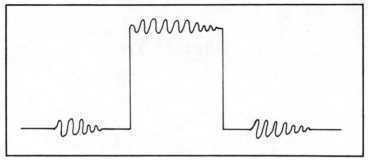

Fig. 25-16. A noisy signal can also cause errors in a digital circuit.

the power supply voltage, a logic 1 as indicated. If the pointer stays near 0, then a logic 0 is being monitored. A voltmeter can also easily locate problematic undefined logic levels (in-between voltages) that can cause erratic circuit operation.

An oscilloscope can be an even more useful tool. It can also measure voltage levels. It can also monitor ac (rapidly changing logic states) signals. Pulse frequency can be checked on an oscilloscope too. Since an oscilloscope directly displays the shape of the input signal, many problems that could cause erratic operation can be spotted. Many of these problems would be difficult to locate by other means.

Figure 25-10 shows what an ideal pulse should look like on an oscilloscope screen. Notice the infinitely steep (straight up and down) rise and fall times. The signal instantly reverses logic states. A practical circuit can't quite produce such an idealized signal. It takes some finite amount of time for the signal to reverse states. This is illustrated in Fig. 25-11. Notice the sloping sides of the waveform. The time it takes the signal to reverse states is called the slew rate. The steeper the slope (lower slew rate), the better the signal.

Sometimes the signal will over-shoot the desired voltage level (see Fig. 25-12), or under-shoot it (Fig. 25-13). In some circuits, the sudden switching of voltage levels can cause some oscillation, called ringing. A pulse with ringing is illustrated in Fig. 25-14. An oscilloscope can also identify distorted (Fig. 25-15) or noisy (Fig. 25-16) waveforms that can also contribute to erratic circuit operation.

Chapter 26

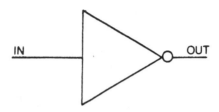

IN ————————————|> OUT

Checking Digital ICs

In troubleshooting a digital electronics circuit, the gates and other ICs are treated like functional "black boxes." Since there is obviously no way to do any internal repairs, we can only check the inputs and outputs and replace the entire chip if anything should be wrong.

Actually, ICs rarely go bad, unless they are physically damaged by heat, or a severe jolt (such as dropping the circuit on a concrete floor). Of course, a power supply or input over-voltage can also damage integrated circuits. And certainly if power is applied to an IC that has been installed backwards, replacement would be called for.

But the chips rarely develop defects in use. Either they work to begin with, and continue doing so (unless forcibly damaged), or they don't. Usually when troubles show up in an operating digital circuit, it is due to something else—e.g., incorrect supply voltages, or bad input signals. But these problems can cause damage to some ICs. And IC failure does occur. This means we must sometimes check ICs.

An IC should also be checked out before it is used in a circuit—especially if it is to be soldered directly, which makes replacement difficult. Sockets can make your life easier. Many experimenters buy ICs from surplus houses. It is definitely a good idea to check grab bag ICs for duds.

Sometimes a surplus digital IC will be partially good. For example three of the gates in a 7400 quad NAND gate package may check out fine, but the fourth is dead. If you have a non-critical circuit that calls for three NAND gates, you may want to try using such a chip. Be warned, however, sometimes the bad gate is caused by an internal crack, or some similar defect that could spread to the other gates within the package after awhile.

One big problem facing many hobbyists when they buy surplus ICs in grab bags, or on PC boards is that all too often these components are not marked with standard type numbers. These devices may be marked with code numbers that are meaningful to the manufacturer, but can't be found in cross reference charts. Sometimes you'll encounter chips with no markings at all. You could just give up and throw out these unmarked ICs. You certainly can't use them if you don't know what the heck they are.

If you're willing to expend a little time and effort, you can determine the functions of many unmarked digital ICs, and save quite a bit of money stocking your parts box. In this chapter we will explore some of the methods for determining the pin functions of unknown chips. Many of these techniques can also be used to check known ICs for potential defects.

FINDING THE VOLTAGE SUPPLY PINS

You can't check out anything else on an IC until you've located its power supply and ground pins. Many digital ICs use a standardized power supply pin arrangement (on 14 pin DIPs, pin 7 is ground and pin 14 is +V, on 16 pin DIPs, pin 8 is ground, and pin 16 is +V). Unfortunately, this logical system is by no means universal, as a glance through the pinout diagrams in the appendices of this book will prove.

If you don't know what the IC type number is, how can you find the correct power supply pins (much less determine the other pin functions)? There are ways. They aren't necessarily easy ways, but the difficulty is more in tedium and the need for close attention to your work, rather than any inherent complexity.

If you are taking your unknown integrated circuits from a surplus PC board, you should be able to find some important clues. If you're very lucky, the boards connection point to the power supply may be marked. In that case, you simply have to trace the trace to the appropriate IC. Occasionally you may find the V+ trace connected to more than one pin on a single IC. This indicates that some of the logic inputs are being held at a constant logic 1 level. We've done this in several circuits throughout this book. It isn't as easy to determine the correct V+ supply pin, but at least you've narrowed down the possibilities.

Similarly, the ground connection may be marked on the board itself. Ground traces on a PC board are often thicker, or have larger "islands" than the other traces. Of course, there should be a ground connection for every IC on the board.

As with the V+ connection, more than one pin on the IC may be connected to ground (forcing some inputs to a constant logic 0 state). At least, the possibilities of the power supply connections have been narrowed down. In addition, you've also determined that the other pins connected to V+ or ground are almost certainly inputs. This will be quite helpful in later stages of the IC identification process.

Even if the V+ and ground lines aren't marked on the PC board, you should be able to find likely candidates by locating traces that go to at least one pin of all of the ICs on the board. Certain polarized components such as electrolytic capacitors and zener diodes can also provide you with some useful clues.

You may have identified the V+ and ground pins already, but in most cases, you will have a few possibilities, but nothing definite yet. This next part of the identification procedure applies to loose ICs, as well as those mounted on printed circuit boards. Remove the IC from the circuit for these tests.

For this next series of tests, you will need an ohmmeter (or a VOM, VTVM, or DMM) set to a medium scale. If possible the ohmmeter test voltage should be fairly low, preferably under three volts. Usually when testing unknown ICs we don't even know the logic family, at least not at first. We certainly don't want to damage the chip with an over-voltage while we are testing it.

Connect the positive ohmmeter lead to pin #1 of the IC, and touch the negative probe to each of the other pins, making a note of the resistance reading in each case. Then move the positive lead to pin #2 and repeat the process. This procedure should be repeated for each of the pins on the IC, until you have made resistance measurements for every pair of pins, in both directions.

Now, look over the values of the resistance readings you have written down. You should notice that one pin produced low readings to all of the other pins when the positive test lead was on this pin. You can reasonably assume that this pin is the chip's ground connection. If the IC came from a PC board, you can double-check this conclusion with your preliminary findings from studying the board's foil traces.

Let's say we found that the ground pin was pin #8. Now look at the readings with pin #8 positive to each of the other pins, and find the lowest reading. This should be the positive voltage supply connection. Once again, you can compare this result with the foil traces, if applicable.

Now, it is time to connect a voltage to the IC and start taking active measurements. Since you do not know the logic family, use the lowest common voltage. RTL chips are powered by 3.6 volts, so this should be your first test voltage. Ideally, you should be working with a variable output power supply with current limiting to avoid damaging the IC if a mistake is made.

In the first phase of the voltage testing, do not apply the power directly to the pins of the IC. Include a 50 mA or 100 mA milliammeter and a 150 ohm resistor in the circuit, as shown in Fig. 26-1. The resistor will help limit the current passing through an IC. Even if there is a dead short through the chip, the maximum current draw will still be a little less than 25 mA. For a good IC, the current reading should be significantly less than this.

If the current reading seems reasonable, remove the resistor from the circuit so you can directly measure the current drawn by the IC itself. If there appears to be no current flow at all (or a very low level) the chip is probably a RTL device. The correct supply voltage is 3.6 volts. You can now move on to locating the chip's

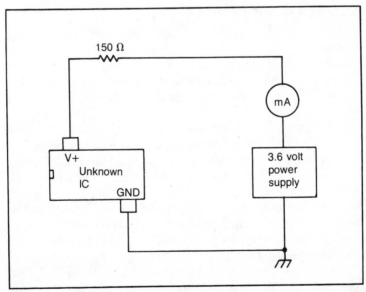

Fig. 26-1. To initially check the current of an unknown IC, you should use this setup.

outputs with the techniques described in the next section of this chapter.

A single TTL gate draws a nominal current of 2 to 4 mA. If your chip draws 12 to 16 mA there is a good chance that it is a quad gate TTL unit. A CMOS gate should draw a much lower current— typically in the 0.1 to 1 mA range. A CMOS quad-gate package will therefore have a total current draw between about 0.4 mA and 5 mA.

Of course, more complex devices (such as flip-flops, counters, decoders, etc.), will probably draw somewhat more current than simple gates, but as a rule-of-thumb, you can usually determine the correct logic family from the current drawn by the device, using the following guidelines:

☐ If the current drawn is less than 0.2 mA assume RTL.
☐ If the current drawn is between 0.4 mA and 10 mA assume CMOS.
☐ If the current drawn is greater than 10 mA measure TTL.

Assuming you've gotten a reading that indicates a TTL or CMOS unit, raise the voltage to 5 volts and recheck the current readings. If the chip is unquestionably a CMOS device, you can try applying a higher voltage, but be very careful at this point.

LOCATING THE CHIP'S OUTPUTS

The next step is to find out which of the pins are outputs. If the IC came from a surplus board, any pins connected to V+ or ground (in addition to the power supply pins themselves) are probably inputs. Using the voltage determined in the previous section to power the IC, measure the voltage from each of the other pins to ground.

A pin voltage greater than about 2.2 volts is probably an output in a logic 1 state. Similarly, a voltage less than 0.3 volts, but greater than 0.0 suggests a logic 0 output. Voltages not fitting within these ranges (i.e., less than 0 volts, or greater than 0.3, but less than 2.2 volts) are probably inputs.

Turn the supply volt on and off several times, while watching the assumed output voltages. A flip-flop output may charge its logic state during this test. A gate will not respond to this test.

Measure the logic 0 currents through a 330 ohm resistor to V+. You should get readings in the 10 to 20 mA range for TTL. The logic 1 currents should be measured through a 330 ohm resistor to ground. Readings from 2 to 30 mA can be expected.

DETERMINING INPUT FUNCTIONS

TTL or DTL inputs will probably float somewhere between 0.8 to 1.8 volts. The current drawn by these inputs will typically be between 0.8 mA and 2.0 mA. The pin to pin resistance readings taken earlier can provide some information as to what is inside the chip. If the resistance between two pins, say pins 4 and 12, is the same in both directions (that is when pin 4 is positive and pin 12 is negative, and when pin 12 is positive and pin 4 is negative), an on chip resistor is indicated between these two pins. A close relationship between the pin functions is implied.

Sometimes the resistance won't be exactly equal in both directions. For instance, you may get a 650 ohm reading with one polarity, and a 740 ohm reading in the other. The two values should be fairly close. If the resistance varies drastically with reversed polarity (say, a few hundred ohms with one polarity, and several thousand ohms when the polarity is reversed), one or more semiconductor junctions exist between the two pins. If the resistance is very high (more than a few hundred thousand ohms) in both directions, you can assume that the two pin's functions are essentially isolated from each other.

One way to find out which pins drive a single gate is to ground each of the input pins one at a time, and measure the voltage from each of the other input pins to ground. A related input should be dragged down along with the grounded input. If there is no effect on another input pin's voltage, the input pins are probably driving different gates.

Now, while monitoring the outputs ground each of the input pins one at a time, and note any changes in the output. Then connect each input pin to V+ and see what effects there are on the outputs. If the results of these tests are inconclusive, or a number of the inputs don't seem to have any effect on any of the outputs, try looking for a chip enable, or inhibit pin. A clock signal may be required. You can manually clock a chip for testing purposes with a SPDT switch, as illustrated in Fig. 26-2. Try grounding or applying V+ to combinations of two or three inputs at a time.

After making a number of such tests, a pattern should start to emerge, and you should be able to make at least a partial truth table. If you have been working carefully, this table should resemble one of the basic types discussed throughout this book.

As an example, let's say our testing has given us the partial truth table shown in Table 26-1. The data we have so far suggests a quad two-input NOR gate. Pins 2 and 3 seem to be the inputs for

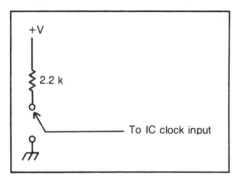

Fig. 26-2. A simple SPDT switch can be used as a manually operated clock.

+V

2.2 k

To IC clock input

output pin 1. Putting a logic 1 on either of these pins results in a logic 1 output. If we ignore the rest of the truth table, we can try to guess and fill in the rest of this truth table.

Right now, the truth table for this gate looks like this:

Inputs		Output
2	3	1
*	*	1
0	*	1
*	0	1
1	*	0
*	1	0

Table 26-1. This Partial Truth Table Is Typical of the Information You Should Find when Checking an Unknown IC.

Inputs								Outputs			
2	3	5	6	8	9	11	12	1	4	10	13
*	*	*	*	*	*	*	*	1	1	1	1
0	*	*	*	*	*	*	*	1	1	1	1
*	0	*	*	*	*	*	*	1	1	1	1
*	*	0	*	*	*	*	*	1	1	1	1
*	*	*	0	*	*	*	*	1	1	1	1
*	*	*	*	0	*	*	*	1	1	1	1
*	*	*	*	*	0	*	*	1	1	1	1
*	*	*	*	*	*	0	*	1	1	1	1
*	*	*	*	*	*	*	0	1	1	1	1
1	*	*	*	*	*	*	*	0	1	1	1
*	1	*	*	*	*	*	*	0	1	1	1
*	*	1	*	*	*	*	*	1	0	1	1
*	*	*	1	*	*	*	*	1	0	1	1
*	*	*	*	1	*	*	*	1	1	0	1
*	*	*	*	*	1	*	*	1	1	0	1
*	*	*	*	*	*	1	*	1	1	1	0
*	*	*	*	*	*	*	1	1	1	1	0

* = floating input—unknow logic level

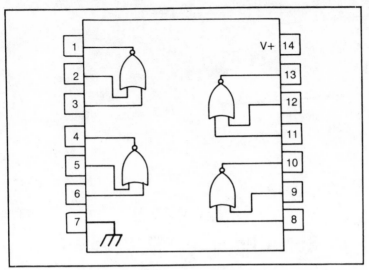

Fig. 26-3. Here is the deduced pinout diagram for the example discussed in the text.

If it is a NOR gate, the complete truth table should look like this:

Inputs		Output
2	3	1
0	0	1
0	1	0
1	0	0
1	1	0

We can try hooking up these input combinations and see if the output responds as predicted.

Once we've determined what's in the IC, we can draw a schematic/pinout diagram of the IC, as shown in Fig. 26-3. You can now use the IC in one of your circuits.

If you want to know more about the chip, you can compare your deduced pinout diagram with those shown in data books, or the pinout diagrams shown in the appendices of this book to learn the device's type number. You should be aware that the appendices are not complete by any means. They illustrate only a few of the most commonly encountered IC chips.

Appendix A
Some Popular TTL Devices

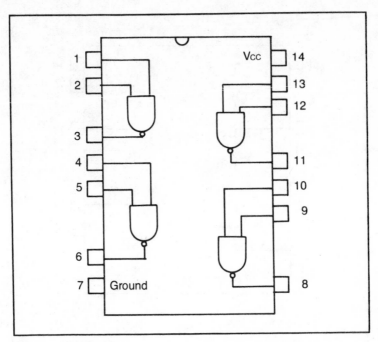

7400 quad NAND gate.

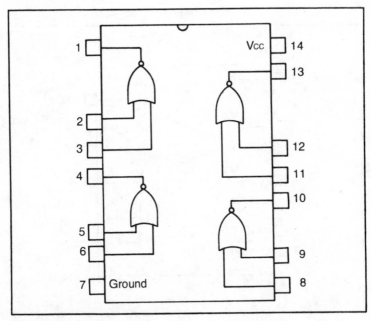

7402 quad NOR gate.

387

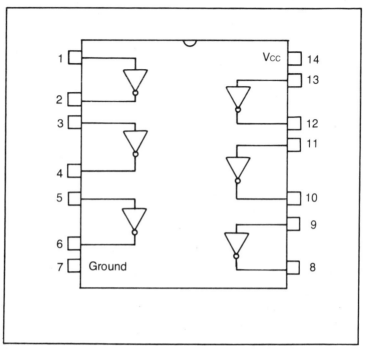

7404 hex inverter.

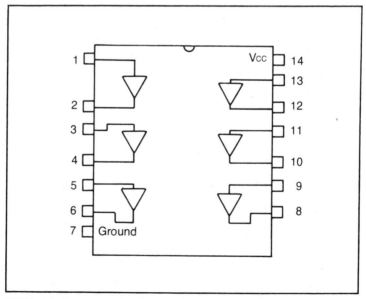

7407 hex buffer.

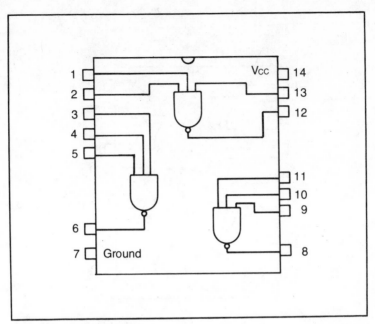

7410 triple 3 input NAND gate.

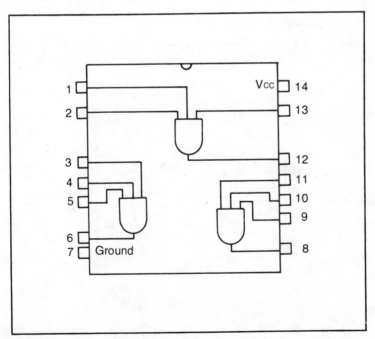

7411 triple 3-input AND gate.

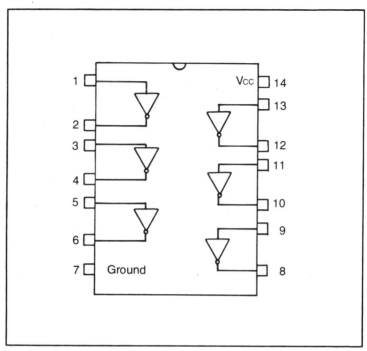

7414 hex Schmitt trigger.

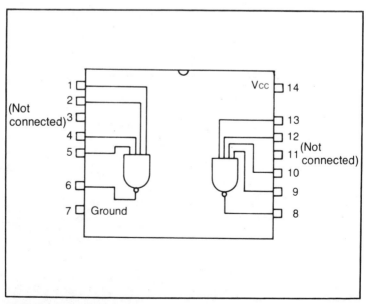

7420 dual 4-input NAND gate.

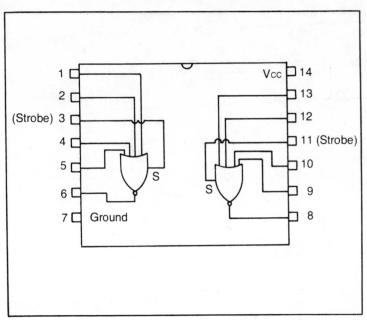

7425 dual 4-input NOR gate with strobe.

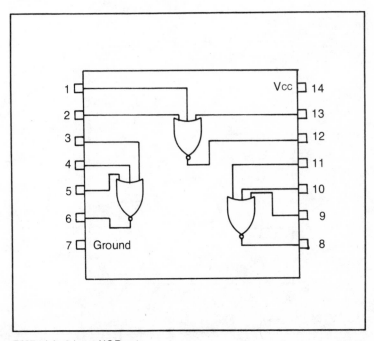

7427 triple 3-input NOR gate.

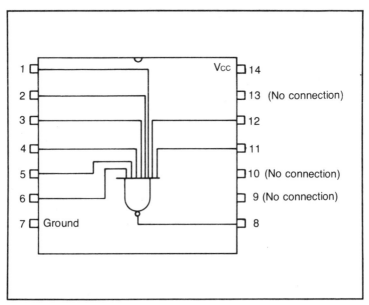

7430 8-input NAND gate.

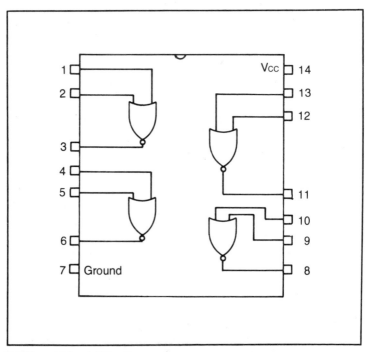

7432 quad 2-input OR gate.

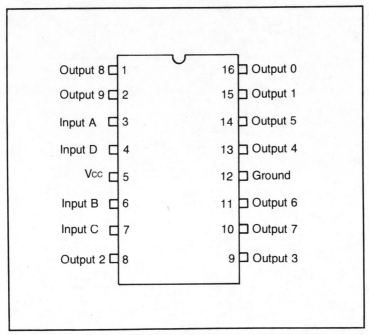

7441 BCD/decimal decoder/driver.

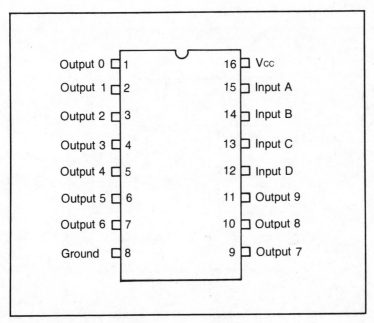

7442 BCD/decimal decoder.

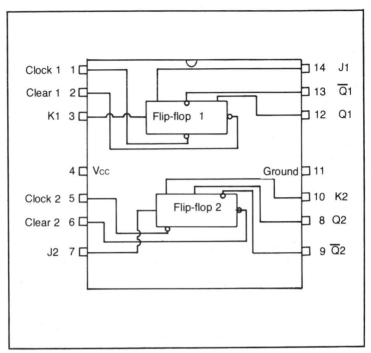

7473 dual J-K flip-flop with clear.

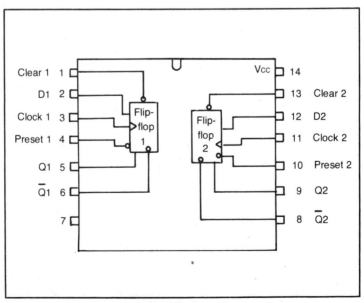

7474 dual D positive-edge triggered flip-flop with preset & clear.

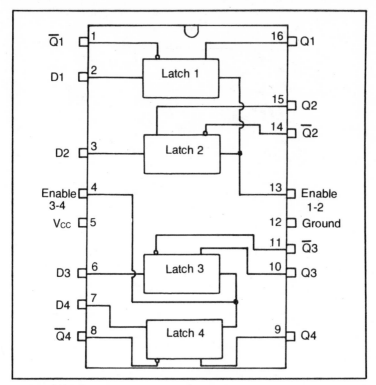

7475 quad latch.

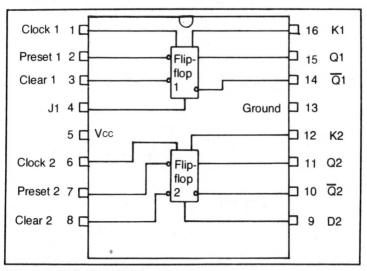

7476 dual J-K flip-flop with preset and clear.

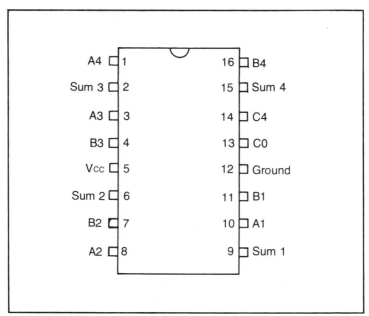

7483 4-bit binary adder with fast carry.

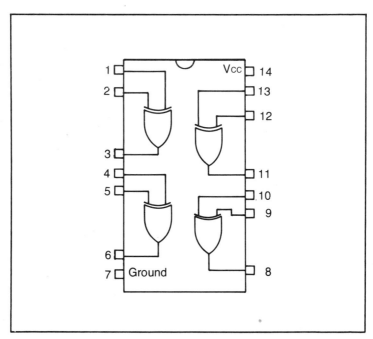

7486 quad 2-input Exclusive-OR gate.

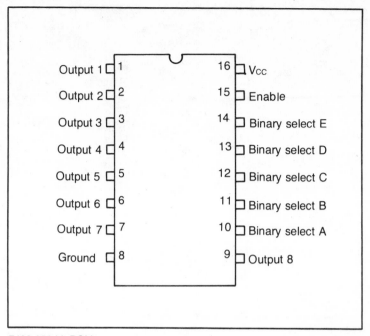

7488 256-bit ROM.

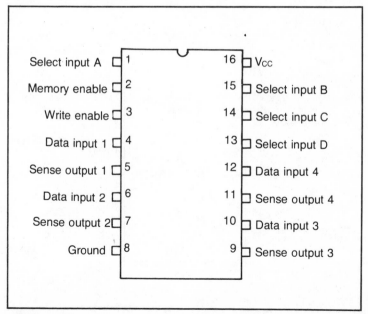

7489 64-bit read/write memory.

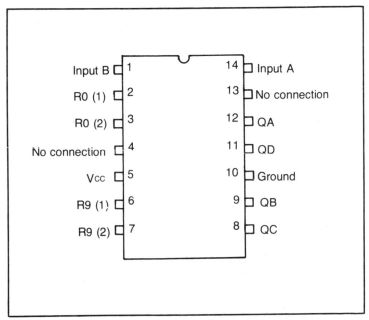

7490 decade, divide-by-12, and binary counter.

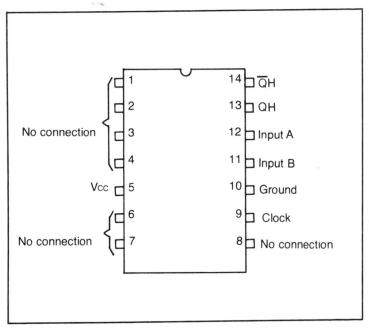

7491 8-bit serial shift register.

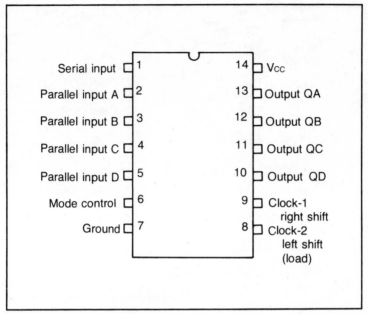

4-bit parallel-access shift register.

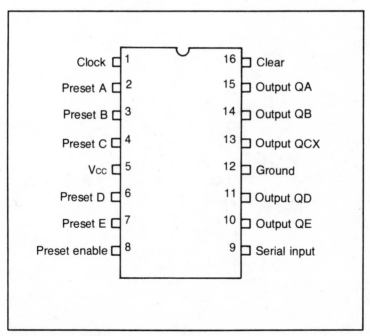

7496 5-bit shift register.

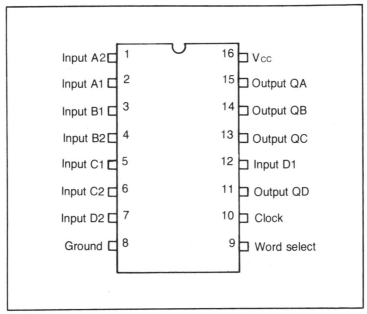

74L98 4-bit storage register.

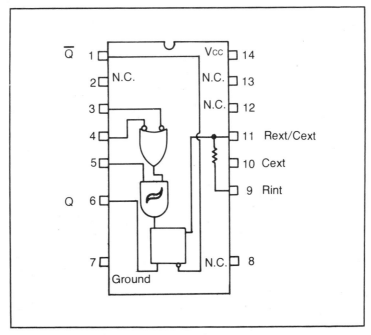

74121 one shot.

400

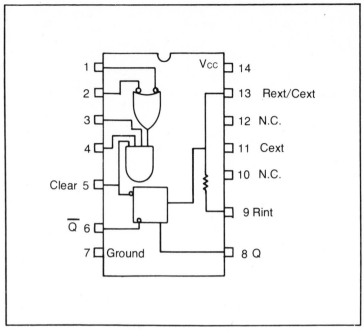

74122 retriggerable one shot with clear.

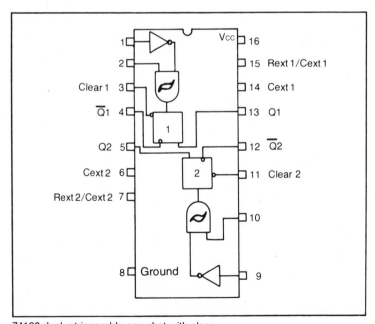

74123 dual retriggerable one shot with clear.

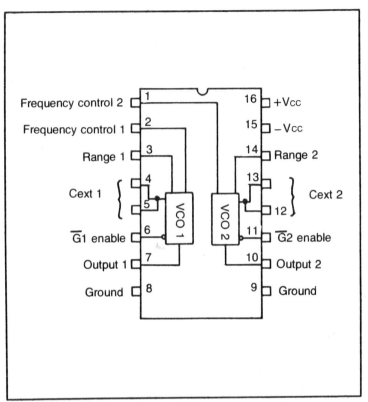

74LS124 dual-voltage controlled oscillator.

Appendix B
Some Popular CMOS Devices

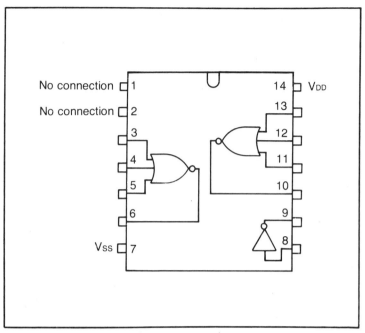

CD4000 dual 3-input NOR gate plus inverter.

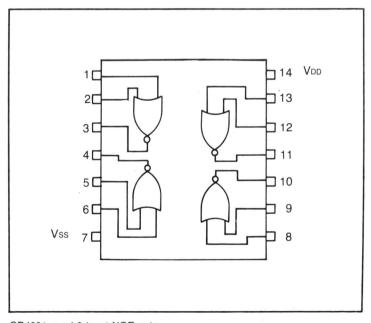

CD4001 quad 2-input NOR gate.

404

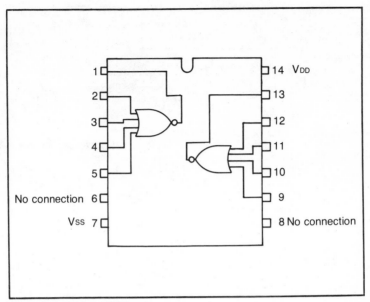

CD4002 dual 4-input NOR gate.

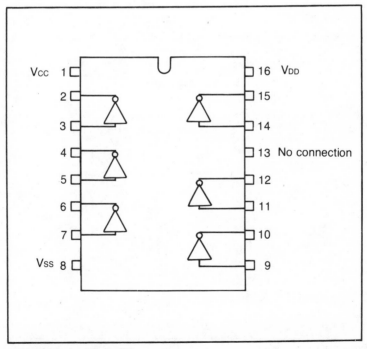

CD4009 hex inverter/buffer.

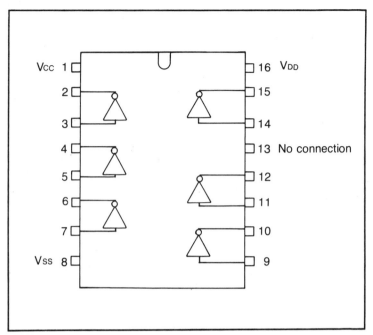

CD4010 hex buffer.

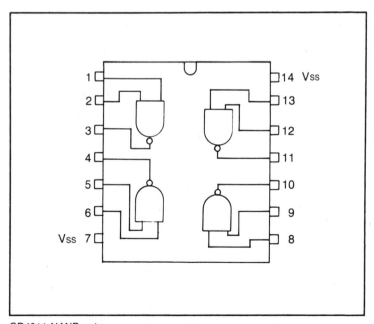

CD4011 NAND gate.

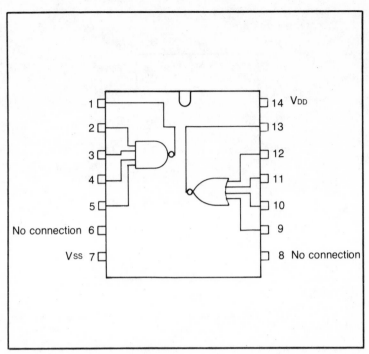

CD4012 dual 4-input NAND gate.

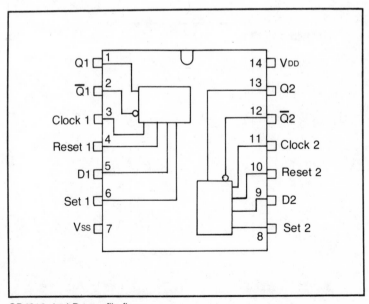

CD4013 dual D-type flip-flop.

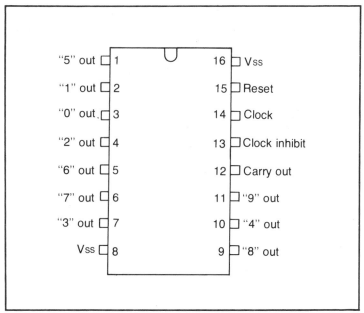

CD4017 decade counter/divider.

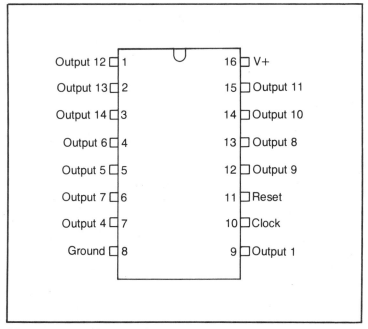

CD4020 14-stage binary counter.

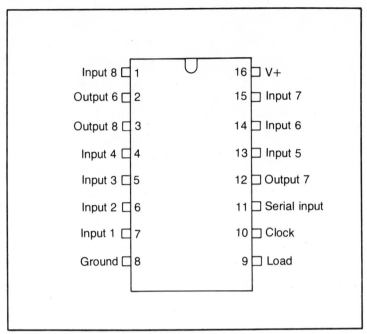

CD4021 8-stage shift register.

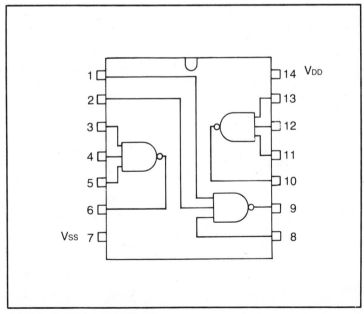

CD4023 triple 3-input NAND gate.

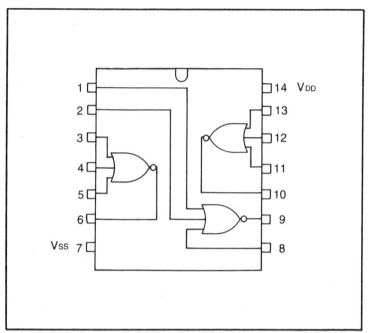

CD4025 triple 3-input NOR gate.

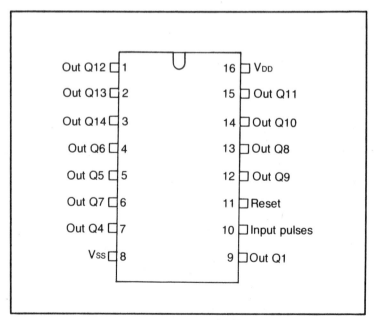

CD4020 14-stage ripple-carry binary counter/divider.

410

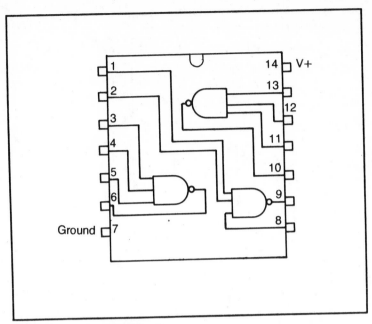

CD4023 triple 3-input NAND gate.

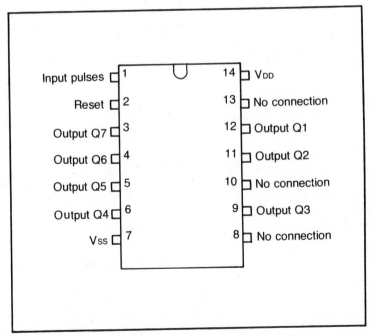

CD4024 7-stage binary counter.

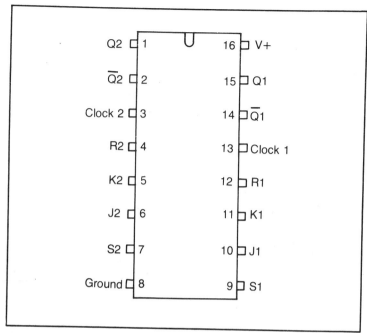

CD4027 dual JK flip-flop.

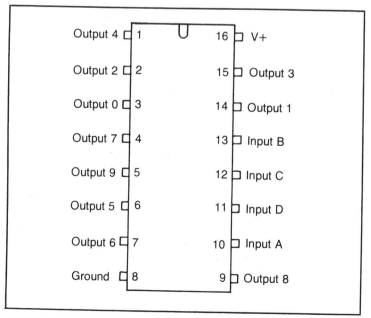

CD4028 BCD-to-decimal decoder.

412

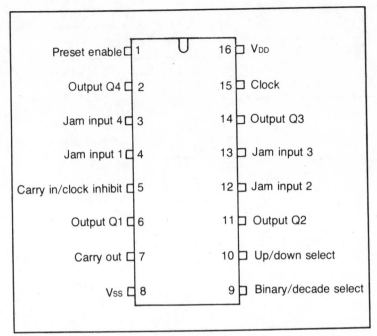

CD4029 presettable up/down counter.

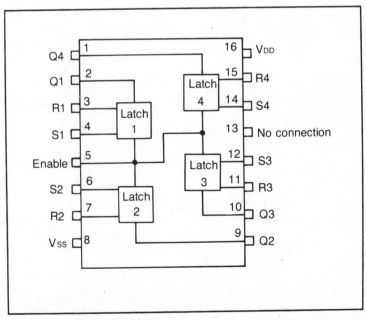

CD4043 quad NOR R/S latch.

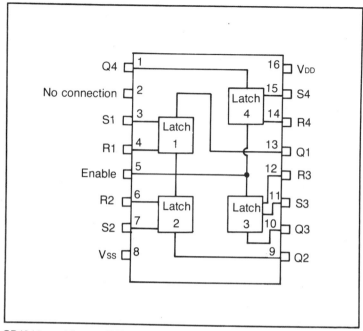

CD4044 quad 3-state R/S latch.

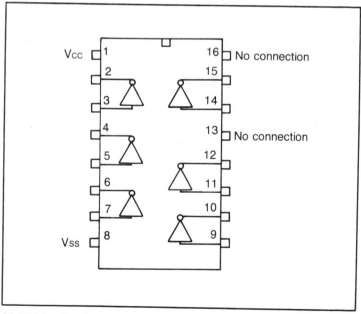

CD4049 hex inverter.

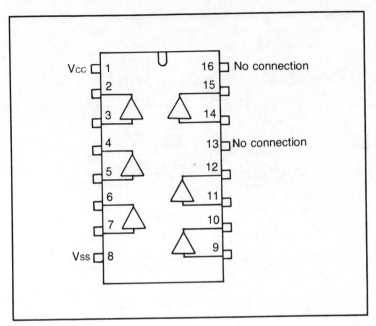

CD4050 hex buffer.

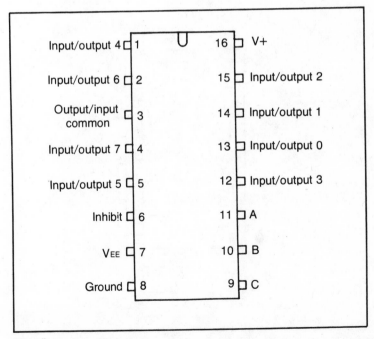

CD4051 analog multiplexer.

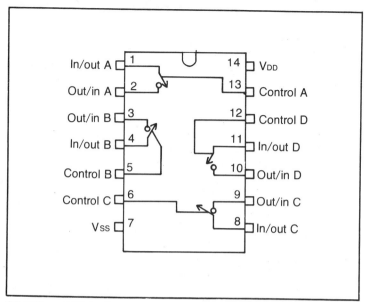

CD4066 quad bilateral switch.

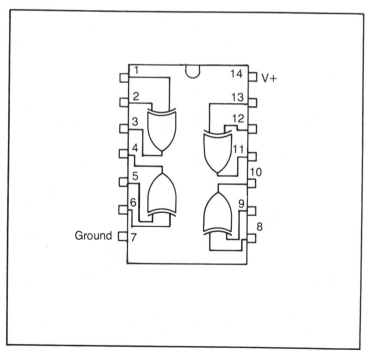

CD4070 quad Exclusive OR gate.

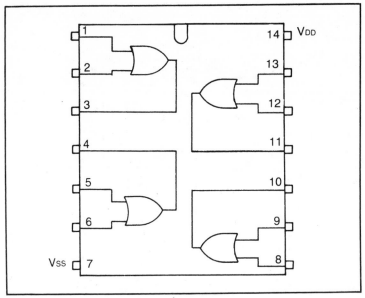

CD4071 quad 2-input OR gate.

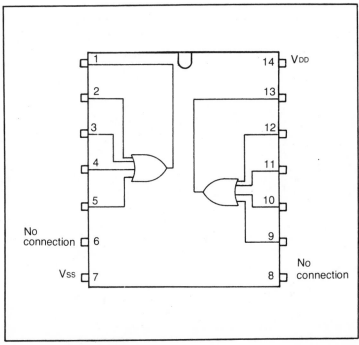

CD4072 dual 4-input OR gate.

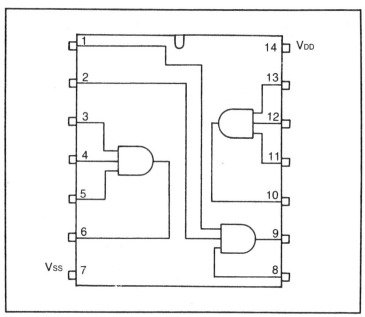

CD4073 triple 3-input AND gate.

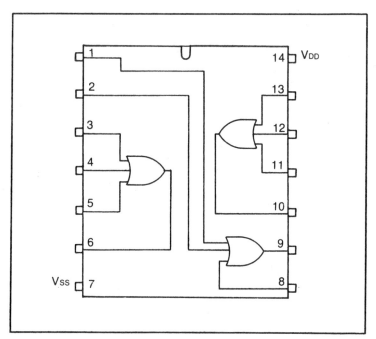

CD4075 triple 3-input OR gate.

418

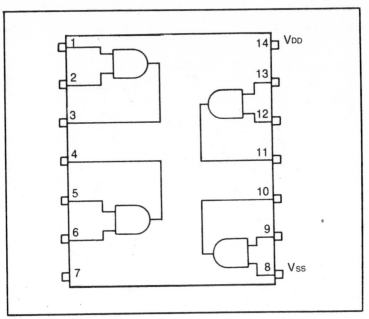

CD4081 quad 2-input AND gate.

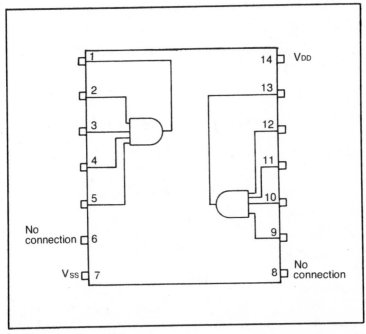

CD4082 dual 4-input AND gate.

419

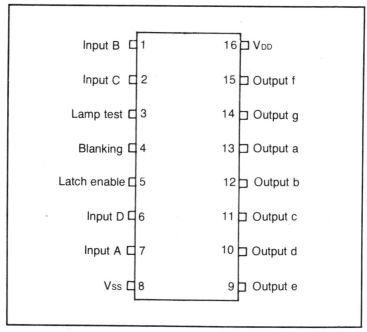

CD4511 BCD-to-7 segment display latch/decoder/driver.

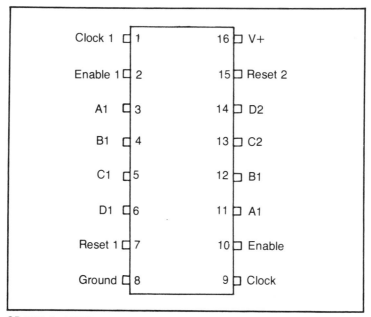

CD4518 dual BCD counter.

420

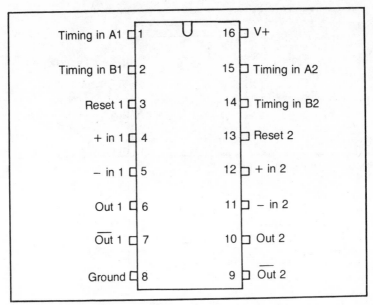

CD4528 dual one shot.

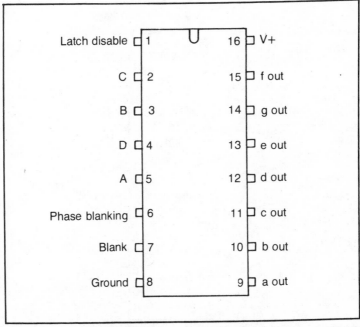

CD 4543 BCD-to-7 segment latch/decoder/driver (sometimes numbered 14543).

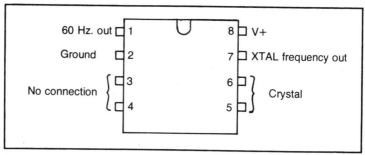

MM5369 60-Hz timebase.

Index

423